Structural Steelwork

Structural Steelwork

Limit state design

A. B. CLARKE
B.Sc., C. Eng, M. I. Struct. E.
Det norske Veritas

and

S. H. COVERMAN
C. Eng, F. I. Struct. E. Consulting Engineer

London New York
CHAPMAN AND HALL

First published in 1987 by
Chapman and Hall Ltd
11 New Fetter Lane, London EC4P 4EE
Published in the USA by
Chapman and Hall
29 West 35th Street, New York NY 10001

© 1987 A. B. Clarke and S. H. Coverman

Printed in Great Britain by J. W. Arrowsmith Ltd, Bristol

ISBN (Hardback) 0 412 29660 8 (Paperback) 0 412 29670 5

British Library Cataloguing in Publication Data

Clarke, A. B.
    Structural steelwork: Limit state design.
    1. Steel, Structural   2. Plastic analysis
    (Theory of structures)
    I. Title   II. Coverman, S. H.
    624.1'821   TA684

    ISBN 0-412-29660-8
    ISBN 0-412-29670-5 Pbk

Library of Congress Cataloging-in-Publication Data

Clarke, A. B. (Antony Bryan), 1947–
    Structural steelwork.
    Includes bibliographies and index.
    1. Steel, Structural.   2. Building, Iron and Steel.
3. Plastic analysis (Theory of structures)
I. Coverman, S. H. (Sidney H.), 1925–
II. Title.
TA684.C58 1987   624.1'821   87–818
ISBN 0-412-29660-8
ISBN 0-412-29670-5 (pbk.)

Contents

Preface

During this century many changes have occurred in the science of structural engineering. Knowledge of structural theory has expanded and the use of computer-aided design has encouraged greater sophistication in the analysis of steel structures – in the elastic and inelastic range.

Steel quality and constructional methods are continually being improved and these factors help in the development of rational design techniques. The objective of this book is to present theoretical and practical methods of designing steel structures based upon current design practice. It is intended for practising engineers, and for engineering students who have an elementary knowledge of theory of structures and strength of materials.

The design examples are generally based upon the recommendations of BS5950: Structural use of steel in building: Part 1, referred to herein as the 'code'. The text includes many illustrative examples to assist the reader in understanding a particular design method. SI units have been used throughout, with the exception of centimetre units, which are used to describe moments of inertia, elastic modulus and radius of gyration. For additional sources of information, a reference list is provided at the end of each chapter.

It is assumed that the reader has obtained a copy of British Standard BS5950: Part 1 for reference when following the worked examples. References to the new code throughout are given generally in the form 'code Table...' or 'code...', quoting the clause number.

The authors wish to acknowledge the various sources from which many of the theoretical principles have been acquired and the cooperation of many organizations, including individuals and institutions, who have granted permission to reproduce their designs, data, tables, graphs and photographs. Whenever possible, the sources and references are noted at the end of each chapter.

Extracts from BS5950 (Parts 1 and 2) (1985) are reproduced by permission of the British Standards Institution. Complete copies can be obtained from BSI at Linford Wood, Milton Keynes, MK14 6LE.

We thank Dharam, S. Soorae, C. Eng. M.I. Struct. E. for checking the text and design examples, and to Maureen Coverman for typing this book.

We would appreciate notification from the readers of any errors found in the text. However, neither the authors nor the publisher can assume responsibility for the results of designers using values and formulae contained in this book, since there are many variables which affect each design problem.

Introduction

DESIGN OBJECTIVES

One of the primary design objectives is that a structure should fulfil its intended function. Besides supporting loads or a combination of loads that occur during its intended life there are other requirements to satisfy such as thermal and sound insulation, water-tightness, fire resistance and foundation settlement. During the intended life of the structure, deterioration takes place owing to the effects of weather, fatigue, ill-use and other causes.

The designer should endeavour to minimize the deterioration of the structure and to ensure its continued serviceability.

Another of the designer's primary objectives is the safety of the structure. The possibility of a collapse must be made remote to avoid loss of life. The achievement of adequate safety is partly analytical and partly a matter of good judgement. Occasionally structural collapses occur while a structure is being erected. Therefore the designer should check that the structure is stable during each erection stage.

Cost of the structure is another important aspect of design. Structural layout and the type of structure selected have a great influence on its cost.

DESIGN PROCEDURES

The principal steps required for a structural design are:

(a) Selection of the layout and type of structure

The layout is usually arranged by the architect in consultation with the engineer. After the basic layout has been established the engineer examines various structural schemes and selects the most economical.

(b) Determination of the type and magnitude of the loads on the structure

The type and magnitude of the loads depends on the location and use of the structure. The loads applied to a structure are:

- Dead load
- Imposed load
- Wind load
- Dynamic load ⎫ special cases
- Earthquake load ⎭

(c) Determination of the internal forces and moments in the structural components

After the magnitude and location of the loads on the structure have been determined, a structural analysis of the structure is carried out to obtain the forces and moments acting on the various structural components, and to check the stability of the structure.

Various load combinations (obtained from the code) are used for the structural analysis to determine (1) the stability of the structure, and (2) the maximum forces and moments acting on the structure.

(d) Material selection, proportioning of the structural members, and the design of connections

The grade of steel to be used in the design needs to be designated.

Member sizes are selected from the Steelwork Design: Guide to BS5950: Part I, 1985, Vol. 1: Section properties: member capacities, published by Constrado. Then calculations are performed to check whether member strength and stability are sufficient to support the applied loads and moments.

After the structural members have been proven adequate, the design of the connections proceeds.

If limit stage design is used in the procedures stated above, then

(e) Check the performance of the structure under service conditions

One of the main serviceability limit states is deflection, and selected members should be checked, using unfactored applied loads, to ensure that the maximum deflections are within the allowable limits.

ELASTIC (ALLOWABLE STRESS) DESIGN

The elastic design method consists of determining the actual or service loads on a structure and then designing the various elements so as not to exceed the allowable stresses. The allowable stresses are based upon the quality of the steel and the actual manner in which it is stressed. The magnitude of allowable stress is a fraction of the yield stress, and the ratio yield stress/allowable stress is the factor of safety. The factors which influence the selection of the factor of safety are:

- Approximations in the method of analysis;
- Quality of workmanship;
- Properties of material which fall below recommended values;
- Stress concentrations and residual stress;
- Deficiency of cross-section dimensions of members;
- Overloading of members.

The allowable stress design is based on an elastic analysis from which moments, shears and axial forces are obtained which the members are designed to carry.

Therefore stresses in members should not exceed the values of the allowable stresses as stated in the Code of Practice, BS449, or Specification used for elastic design.

The imposed loads acting on a structure are obtained from BS6399: Part I: Code of practice for dead and imposed loads (formerly CP3: Chapter V: Part I). Although the values of these loads have been in use for many years, they are mainly empirical in nature, being derived from observation and assessment. There are possible inaccuracies in the loading in addition to other indeterminate factors which are allowed for by using permissible stresses well below the strength of the material. Therefore, with various load combinations, the factor of safety may be very high.

This fact has been recognized in the presentation of recent codes, in which a reduction in live load when a member carries a large area, or many floors, or the increase in allowable stresses when considering the effect of wind or earthquake loads, is permitted.

Thus there is difficulty in evaluating the actual capacity of a structure to resist the loads acting upon it.

Limit state design gives a clearer assessment of the margins of safety used to cover the uncertainty of loading and other indeterminate factors. A brief description of this method of design follows.

LIMIT STATE (LOAD FACTOR) DESIGN

Limit state design is a procedure in which consideration is given to various and distinct performance requirements. These requirements correspond to limiting conditions in which the structure becomes unfit for its intended use.

The limit states generally considered in structural design are:

- Ultimate (safety) limit states
- Serviceability limit states

For structural steelwork the relevant limit states are:

(a) Ultimate

- Strength limits (yielding, rupture etc.)
- Overturning (loss of equilibrium)
- Elastic or plastic instability
- Fatigue or brittle fracture

(b) Serviceability

- Excessive deformation
- Excessive vibration
- Corrosion and durability

For the limit states where the load level is the criterion (e.g. strength or stability) the 'service' or 'actual' loads are multiplied by load factors and the products are generally referred to as 'factored loads' or 'ultimate loads'.

A load factor of unity is used for the serviceability limit state. Thus the design load used in a deflection calculation is the actual specified 'service' load.

Partial safety factors

The overall factor of safety embraces a number of considerations of which the main ones are:

- Variation in the loading (loads may be greater than envisaged)
- Effects of load combinations (e.g. dead load plus wind, for overturning check)
- The difference in the strength of the steel from that assumed
- Variation in dimensions of members
- Design assumptions used

The above effects have been combined into three partial factors:

γ_l Loading factor
γ_p Structure performance factor
γ_m Materials factor

BS5950 gives values for γ_f which represent the product of γ_l and γ_p.

The γ_f factors are applied to the appropriate 'service' loads to obtain the 'factored' loads.

The γ_m factor is incorporated into the design strength of the steel which, in BS5950, is given as $1.0 \times$ minimum yield strength but it should not exceed $0.84 \times$ minimum ultimate tensile strength of the material.

The code tabulates the design strengths of various grades of steel.

Load factor values

The values for γ_f to be used for design are shown in the table below.

The minimum γ_f factor, 1.0 (from Table 1.0) for dead load is only used to counteract the effects of other loads, e.g. combination of dead load plus wind load, where the factored overturning moments should not exceed the minimum restoring moment.

Where two types of non-permanent load occur (such as imposed and wind loading), then each is first considered separately (with the dead load) using their maximum γ_f factor.

Table 1.0 Load factors

Type of load or combination of loads		γ_f
Dead load:		1.4
	Countering overturning or uplift	1.0
	Acting with wind and imposed loads combined	1.2
Imposed load:	In the absence of wind load	1.6
	Acting with wind load	1.2
Wind load:	Acting with dead load only	1.4
	Acting with imposed load	1.2
	Acting with crane loads	1.2
Overhead travelling crane loads:		
Vertical or horizontal crane load (considered separately)		1.6
Vertical and horizontal crane load acting together		1.4
Crane loads acting with wind load		1.2
Forces due to temperature effects:		1.2

Note: For crane loading with wind or imposed load the value of γ_f for dead load may be taken as 1.2. Table 2 – BS5950: by kind permission of the British Standards Institution.

The numerical values given in Table 1.0 for different types of loading are selected to give a uniform probability of safety.

Strength limit

The safety limit state of strength is the maximum load that can be carried.

The elastic or plastic theory may be used to predict the safety limit, and either of these methods may be applied to analyse simple statically determinate structures or continuous construction.

The structural theory most appropriate to the potential failure mode should be used for the design of individual members or connections. Often, alternative types of failure require checking.

Whether the overall structural analysis is elastic or plastic, for failure by general yielding, the moment carrying capacity is the plastic moment. (See Chapter 5 for methods of plastic design.) Failure may occur at a lower load level than that required to obtain the plastic moment, due to either local or overall buckling. Therefore, both the moment capacity and the buckling resistance moment require checking.

The main advantages of limit state design are a more uniform factor of safety, and an appreciation of structural behaviour, as the designer needs to check safety and serviceability requirements for a number of limit states. Strength, stability and deflection are the most important limit states, and each set of calculations is compiled for a particular performance requirement. It remains to be seen whether the partial factors stated in the code (see Table 1.0) lead to greater economy of construction.

REFERENCES

1. Mitchell, G. R. and Woodgate, R. W. (1971) *Floor Loadings in Office Buildings – the Results of a Survey.* BRS Current Paper 3/71.
2. CIRIA (1977) *Rationalisation of Safety and Serviceability Factors in Structural Codes.* Report No. 63.
3. CIRIA (1972) *Variability in the Strength of Structural Steel – a Study in Structural Safety. Part 1. Material Variability.* CIRIA Technical Note 44.
4. Baker, M. J. and Taylor, J. C. (1978) *The Background to the New British Standard for Structural Steelwork. Application of Limit State Design to the Code.* Imperial College, London.
5. *Aims of Structural Design* (1969) A Report of the Institution of Structural Engineers.
6. Bates, W. (1976) *Limit State Design – A Review of Limit State Design Concept and its Applications to the Design of Structural Steelwork,* BCSA.
7. CEB-CEMC-CIB-FIP-IAB Joint Committee on Structural Safety (1976) *First Order Reliability Concepts for Design Codes.* CEB Bulletin No. 112.
8. Lambert Tall (1974). *Structural Steel Design,* Wiley, New York.
9. Bresler, B., Lin, T. Y. and Scalzi, J.B. (1968) *Design of Steel Structures.* Wiley, New York.
10. BS5950. Part 1. 1985 Structural use of Steelwork in building, Part 1. Code of practice for design in simple and continuous construction: hot rolled sections.

Metallurgical considerations 1

1.1 INTRODUCTION

It is possible for steel to fail at a stress below the nominal yield values stated by the manufacturer. The reasons for such failure, and methods of avoiding it, are briefly stated in this chapter.

Apart from the more unusual forms of failure (such as fatigue failure) the most common serviceability failure is corrosion. The steelwork designer must therefore identify the atmospheric/chemical environment of the location of the steelwork structure, and select the protection system accordingly. Some important structural projects have proved inordinately expensive because of the untimely arrival of corrosion, due to the selection of an inadequate painting system. The subject of painting is treated in depth in the publications [1–3] listed at the end of this chapter. In general there is no single protection system that is suitable for all locations, but for locations that could be classed between 'severe' and 'mild' the Building Research Establishment [1] lists suitable treatment for most types of steelwork.

Steel is not a homogeneous material; all steels include microscopic impurities which tend to be preferentially orientated in the direction of mill rolling. Some important impurities also tend to stay near the centre of the rolled item, due to their preferential solubility in the liquid metal during solidification, that is to say near the centre of rolled plate, and at the junction of flange and web in rolled sections. The main body of steel consists of a cohesive mass of crystalline domains whose character is determined by many factors, but most noticeably (from a structural engineer's point of view) the steel microstructure is affected by the rate of cooling from elevated temperatures. Faster cooling will result in smaller crystal grain sizes, generally resulting in some increase in strength and toughness.

The resistance of steel to fatigue, brittle fracture at low temperatures, and the effects caused by welding should be considered at the inception of design. For many steel structures the factors that could cause problems related to the above will not apply, but the designer should be aware of these factors in order to eliminate them. A consultant metallurgist may be required in order to specify the chemical content of the steel, mode of manufacture and testing, and advise on suitable welding processes. The role of the metallurgist is particularly important in the engineering of large off-shore steel platforms, necessitating close co-operation between the structural engineer and the metallurgist.

The designer finally has a responsibility to ensure that the structure will remain structurally stable for its entire design life (within the loading extremes stated in his design). This can occasionally mean that a structure is kept under periodic surveillance in order to prevent the growth of crack defects that may shorten the lifespan of the structure, or initiate some form of collapse. This form of checking is described in Section 1.4.

1.2 STEEL CHARACTERISTICS

1.2.1 Selection of a suitable steel

For the majority of structural engineering purposes, steel of a recognized standard should be used, such as BS4360, the related international standard ISO630, or the American ASTM standards. Selection of steel type may depend on the strength required in bending, the 'ductility' of the steel (i.e. the ability to carry a load above yield stress coupled with its subsequent failure in a plastic manner) and the service temperature of the surroundings. Impurities and crystals lie approximately parallel to the plate face which results in the fact that most steels lack ductility in the Z thru' thickness direction, that is, normal to the plane of rolling, and consequently cannot yield by small amounts (as they are free to do in the other two axes). This feature is of concern in heavy connections under high load, and is discussed with related difficulties in the section on lamellar tearing. If service temperature or ductility requirements are not met by steel listed in the standards, then the advice of a metallurgist should be sought. The steel that the metallurgist selects should then be tested for its resistance to brittle fracture before the steel order is placed. A description of the factors affecting brittle fracture and a method of dealing with the problem are given below.

1.2.2 Brittle fracture

Brittle fracture of steel is most likely to occur where there is triaxial stress (sometimes called hydrostatic stress). In fact, any *solid* material can be made to fail in a brittle fashion by application of a suitable triaxial loading system. In practice brittle fracture is normally associated with wide plates or 'heavy' steelwork in low temperatures. Heavy steelwork can be considered more at risk from this type of failure because of the higher degree of restraint provided by thick plates, wide plates, and deep webs, which tends to make loading more truly tri-axial. Lowering of temperature causes the steel to become 'brittle' which assists the propagation of any existing crack or defect. The crack itself is propagated by triaxial stress near (ahead of) the crack tip. A graphical indication of brittle failure is given by typical Charpy test results (Fig. 1.1) where the energy absorption of mild steel can be seen to drop off below $-5°$ C.

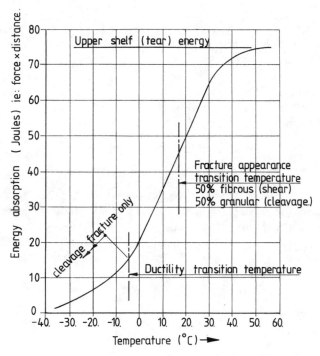

Fig. 1.1 Typical Charpy 'V' notch graph for mild steel.

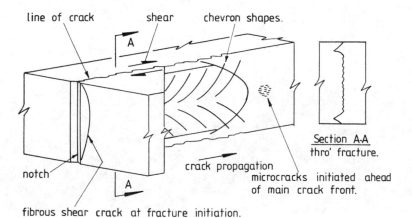

Fig. 1.2 Brittle fracture; crack propagation from a notch.

The factors affecting brittle fracture strength are as follows:

(1) Steel composition, including grain size of microscopic steel structure, and the steel temperature history
(2) Temperature of the steel in service
(3) Plate thickness of the steel
(4) Steel strain history (cold working, fatigue etc.)
(5) Rate of strain in service (speed of loading)
(6) Internal stress due to welding contraction
(7) High applied stress
(8) Flaw size and type of flaw (i.e. planar, profile and volumetric).

In general slow cooling of the steel causes grain growth and a reduction in the steel toughness, increasing the possibility of brittle fracture. Residual rolling stresses marginally reduce the fracture strength, whilst service temperature governs the tensile stress at which the steel will fail in brittle fracture.

The usual appearance of a brittle fracture crack is sketched in Fig. 1.2, see also Fig. 1.3. The occurrence of a brittle fracture is generally instantaneous.

There are many types of tests to determine the risk of brittle fracture. Three of these are:

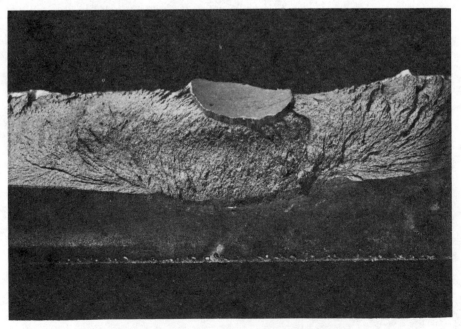

Fig. 1.3 Brittle fracture surface. (Courtesy of The Welding Institute.)

(1) KIc test
(2) CTOD test
(3) Charpy V-notch test.

The first two are large-scale laboratory tests on full plate thickness samples and the third is a small scale laboratory test on a small machined sample.

Tests 1 and 2

The study of the behaviour of cracks in loaded material is termed fracture mechanics. The KIc test was developed by the Americans and forms the basis of fracture mechanics study. Historically speaking, KIc was one of three test types on thick plate, but the passage of time has shown KIc to be the most significant (see Fig. 1.4). KIc represents the study of crack growth in elastic plane strain conditions.

When these plane strain conditions break down and plastic flow starts to occur at the crack tip the measurement of Crack Tip Opening Displacement (CTOD) becomes of primary interest. The KIc and CTOD tests are fully compatible so that results for specimens can be recorded for the relevant fracture type. A defect in the form of a carefully introduced notch is usually placed in a region of interest in order that an assessment may be made of the significance of cracks occurring in service conditions. Using the data obtained in the tests, a critical defect size in the existing structure, detected by non-destructive testing, may be checked for acceptability using methods indicated in British Standard PD6493 (which tends to give conservative results because of built-in safety factors).

Test 3

The Charpy V-notch test to BS131: Part 2: 1972 (Fig. 1.5) is widely used to obtain individual acceptable absorbed energy values for steels, by the use of small notched specimens. It is an empirical test, but the values obtained acquire their significance for end-use simply through the accumulation of past experience. The test is popular because of its low cost and quick results.

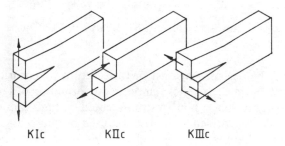

KIc   KIIc   KIIIc

Fig. 1.4 Crack modes (after [10]).

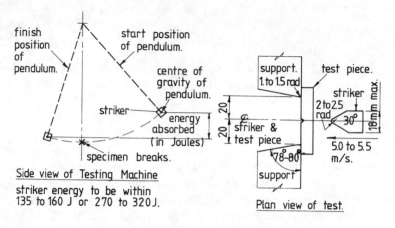

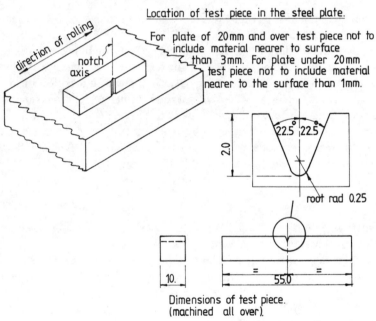

Fig. 1.5 The Charpy V-notch test (from BS131:Part 2:1972). (Courtesy of the British Steel Institute.)

Steelwork can be designed against brittle fracture by ensuring that welded joints impart low restraint to plate. Temporary deformation of a minor type may be allowed to occur, since high restraint could initiate failure. Stress concentrations should be avoided – typically caused by sharp angular changes in direction, 'abrupt' changes in shape and in holes. In service conditions, defects classed as 'acceptable' could be increased to a size that could fail by brittle fracture because of the presence of fatigue or corrosion fatigue conditions. The manner in which the structural engineer approaches the problem of fatigue is described in the next section.

1.2.3 Fatigue strength

The application of cyclic load to a structural member or connection can result in failure much lower than yield. Fatigue calculations are usually carried out for the design of structures such as railway bridges, supports for large rotating equipment and supports for large open structures subject to wind oscillation.

The usual appearance of the face of a fatigue fracture is sketched in Fig. 1.6, see also Fig. 1.7. The fracture line is not easily visible in service conditions but can be detected with non-destructive testing techniques.

Governing criteria

Factors adversely affecting the fatigue resistance of a member or connection may be listed as follows:

(1) Stress concentrations
(2) Residual stresses in the steel
(3) Work hardening due to only moderate cold rolling
(4) Welding causing shrinkage strains
(5) Nature of cyclic loading – i.e. number of cycles for each stress range

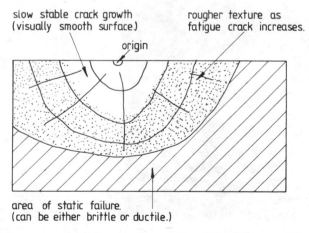

Fig. 1.6 Appearance of a fatigue fracture surface (after Fisher and Struik [21]).

Fig. 1.7 Fatigue fracture. (Courtesy of The Welding Institute.)

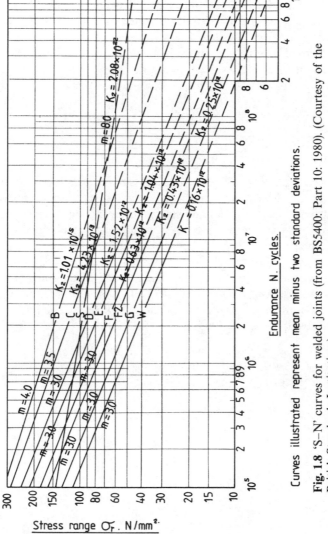

Curves illustrated represent mean minus two standard deviations.

Fig. 1.8 'S–N' curves for welded joints (from BS5400: Part 10: 1980). (Courtesy of the British Standards Institution.)

(6) Service temperature of the steel

(7) Environment (important in the case of corrosion fatigue).

The weakest parts of a steel structure from a fatigue point of view are the connections. For bolted joints HSFG waisted bolts (see Section 3.3) have been specifically designed for fatigue situations as they resist direct tension by means of the prestress in the bolts. These bolts are often used in locations where welding would be impractical, such as machine bases subject to dynamic loading, or the end connections of large flare booms subjected to wind oscillation. Welded joints, which are classified in BS5400:Part10, can be checked for stress by a computer program and compared with the graphical data provided in the British Standard (Fig. 1.8).

S–N curves

Most of the commonly used welded steel joints have been tested to fatigue failure and the results plotted as shown in Fig. 1.9. The BS5400:Part 10 classification for joints subjected to fatigue (including stress concentration factors), together with design curves, enable the designer to computerize calculations on structural fatigue problems.

The design curves are straight lines tangential to experimental failure curves, corresponding to a 97.7% survival limit. The value of S is the sum of the maximum tensile and compressive stresses per cycle, called the 'stress range' and N equals the maximum number of cycles of S in the design life of the joint (Fig. 1.8).

The relationship between S and N is of the form:

$$\log_{10}N = \log_{10}K_2 - m\log_{10}\sigma_r$$

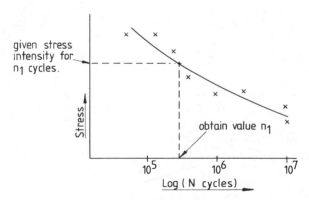

Fig. 1.9 'S–N' curve for a member or connection.

where N = number of cycles

$\quad \sigma_r$ = given stress

$\quad K_2$ = a value dependent on joint detail (Table 8, BS5400)

$\quad m$ = slope of graph and curves show d standard deviation below
the mean.

($d = 1$ corresponds to approx. 15.9% failure probability; $d = 2$ corresponds to approx. 2.3% failure probability.)

For a joint subjected to a loading spectrum (a number of loading repetitions nx of several stress ranges Sx) the number of cycles to failure should be determined in each case from Fig. 1.8 and summated as follows:

$$\sum \frac{nx}{Nx} \ngtr 1.0$$

$$\left(\frac{n_1}{N_1} + \frac{n_2}{N_2} + \cdots \frac{n_n}{N_n} \right) \ngtr 1.0 \text{ (Miner–Palmgren summation)}$$

where N_1, N_2, N_n = No. of cycles to failure at given stress, n_1, n_2, n_n = No. of actual cycles at given stress.

When considering a single stress range, fatigue need not be considered at stresses (S_0) below those corresponding to $N = 10^7$ (Fig. 1.8) where S_0 is the stress below which a crack would not be propagated. However, a loading *spectrum* is taken to a value of $N = 2 \times 10^7$ because high stress could propagate cracks which may be propagated further by low stresses. These values are for joints in air; for sea water the cut-off value is taken as 2×10^8 cycles.

1.2.4 Stress corrosion and corrosion fatigue

Stress corrosion is the name given to the phenomena where cracks in material subjected to sustained tensile stress in a corrosive environment undergo chemical attack at the crack tip. The main area of interest for the structural engineer is the stress corrosion of offshore structures and special structures exposed to aggressive chemical surroundings. Investigations of stress corrosion cracking are generally directed towards obtaining suitable steel and plate details, in order that crack growth rates are slow enough to be detected during periodic inspection. Corrosion fatigue is a related problem where fatigue conditions occur in the aggressive environment. Corrosion takes place at the tip of a crack because this region is anodic to the rest of the exposed surface. Corrosion products can often wedge the crack open and atomic hydrogen released by corrosion reaction can dissolve in the metal, sometimes causing embrittlement. A metallurgist should advise on the solution to any particular problem. Present day methods of dealing with both stress corrosion and corrosion fatigue include particular care in the choice of material, the design, and fabrication technique, and often the use of cathodic protection. Due attention must be paid to the correct range of potential for cathodic protection. Too low a potential leaves the material susceptible to corrosion fatigue. Too high a potential allows hydrogen to enter, with the risk of hydrogen embrittlement if the yield stress is sufficiently high.

1.2.5 Design against fatigue

Information on the loading spectrum should be obtained, based on research or documented data. If this information is not readily available, then assumptions must be made with regard to the nature of the cyclic loading, based on the design life of the structure. If stress histories are available, the reservoir method of cycle counting should be used, as noted in BS5400, appendix B. The connection should be classified into a minimum number of groups; and for bolted connections, HSFG 'waisted' bolts should be employed. Welded connections may be classified according to the procedure outlined in BS5400, and fatigue failure designed against by use of the S–N curves, as previously described.

1.3 STRUCTURAL EFFECTS OF WELDING

1.3.1 General

Welding is a principal method of joining steelwork members. The join is generally made by causing an electrical arc at the end of a filler rod (called the *electrode*) in the region to be welded. A 'flux' is usually added either as a coating to the electrode or as a powder, as in the automatic submerged arc process, where the arc is hidden in a continuous pile of flux. The flux has several functions, two of which are: to release gases to protect the cooling weld from harmful gas in the atmosphere, and to precipitate unwanted compounds in the form of surface slag. The welding of steel creates the possibility of damage such as local defects in or adjacent to the weld, and distortion of the structure due to shrinkage of structure and weld caused by heating.

1.3.2 Local defects in the weld

Defects not already present in the steel material may be introduced by welding. These may be listed briefly as porosity, undercut, lack of fusion or penetration, slag inclusions, and change in the microstructure of the parent metal immediately adjacent to a weld (called the Heat Affected Zone, HAZ). *Porosity* is associated with the beginning of manual metal arc welds where air has been trapped by a longer arc than normal. The defects can be remedied at the time of welding by travelling back over the first 15mm of weld. *Undercut* is a surface defect, where the parent metal is melted back and is not filled with weld metal. It can be caused by use of the wrong electrode, travel speed or wrong electrode angle.

Lack of fusion between weld metal and parent metal can be caused by use of a lower welding current than necessary, or again, wrong electrode angle. *Slag inclusions* are non-metallic materials trapped in the weld metal due to insufficient care in making several weld passes, and failing to remove slag from a previous weld run. All of the above defects (except for change of microstructure in the HAZ) can be largely avoided by ensuring that the material to be welded is absolutely clean, dry, free from mill scale, and that the equipment is used in the correct manner. Repairs to defects in the weld region usually consist of gouging out the affected area and filling in several weld runs, using a weld of low hydrogen content and lower tensile strength.

1.3.3 Welding distortion

When steelwork is welded, distortion can appear due to shrinkage of both the weld and the parent metal. The reason for the distortion is that the heated steel tends to expand in the direction of least restraint (usually the plate thickness) but shrinks in all directions on cooling, causing tensile stress in the plate which can manifest itself in distortion. The weld itself shrinks in both the longitudinal and the transverse direction, pulling the plate with it.

In general, shrinkage occurs very shortly after the weld is laid and before the welding run is complete, so that the joining of two plates by a butt weld can be made more difficult by distortion pulling the plates together ahead of the welding. This difficulty may be overcome by tapering the gaps between short plates by experiment and on longer joins by staggering the weld lines and welding as shown in Fig. 1.10. This 'shrinkage' of steel makes all fabrication work more difficult in that the effect of shrinkage must be allowed for, so that a slightly oversize fabrication when welded becomes dimensionally correct, or possible to correct by machining.

Temporary attachments are often used to maintain the fabrication shape, such as stiffeners, bracing or strongbacks (see Fig. 1.11). Presetting and prebending are techniques used to correct distortion by anticipating the amount of distortion and setting plate at the opposite angle. Plate girder flanges are

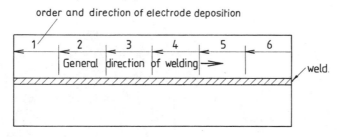

Fig. 1.10 The back-step welding method of reducing distortion.

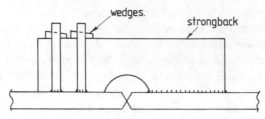

Fig. 1.11 The use of a strongback to prevent plate movement one way.

often prebent outwards to counteract welding shrinkage which tends to pull the flanges in towards each other.

Post-welding correction of distortion can be achieved by the judicious use of local heating. For instance, methods similar to that used to precamber beams by heating in triangular 'wedges' may also be used to correct members in truss assemblies (Fig. 1.12). In box girder construction, stiffeners provide restraint against some distortion, but it is common for the panels to buckle. Careful heating of the panel centre in spot areas in a regular pattern can be used to pull the plate flat.

The general rules to follow in counteracting fabrication distortion are to use the smallest weld sizes possible, ensure good fit up of joints (correct gap for weld with adjustment for shrinkage), balance welds about the neutral axis, work from the centre of the fabrication outwards, and avoid welding all the members at a node at one time. Further guidance on the subject is given in reference 26.

1.3.4 Cracking

Solidification cracking

This type of cracking can take place if shrinkage of the fabrication occurs before the weld has solidified enough to carry the force. It is often associated with 'T' type connections between two plates using full penetration welds in two runs (one each side) instead of four smaller runs (two each side) which is preferable because of the lower amount of weld dilution that takes place.

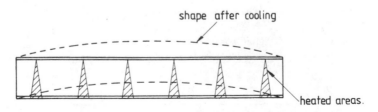

Fig. 1.12 Precambering of a beam by heating.

HAZ cracking

When moisture is present in the vicinity of a weld, the heat of welding drives off the oxygen but the hydrogen tends to disperse in the HAZ. If the cooling rate of the weld is slow, the plate will tend to release the hydrogen to the atmosphere. Preheat would assist the slow cooling. Faster cooling causes a mechanism whereby the presence of the hydrogen may cause a crack in the HAZ. This type of crack generally appears several hours after a weld has been laid. It is therefore usual to delay NDT methods for about 48 hours to allow time for the defect to occur.

1.3.5 Lamellar tearing

Lamellar tearing may be defined as the separation of parent metal (roughly parallel to the plate surface) caused by through-thickness strain induced by weld shrinkage. It is more likely to occur in large connections with heavy members under high stress. A lamellar tear starts in the HAZ, tending to occur in planes normal to the applied load. Generally, when using British steel, only plate over 20mm thick need be considered for lamellar tearing.

The tear is step-like in appearance, having a fibrous 'woody' texture. These two characteristics typify a lamellar tear as opposed to a hydrogen caused crack. Figure 1.13 gives an indication of the appearance and a typical location, see also Fig. 1.14.

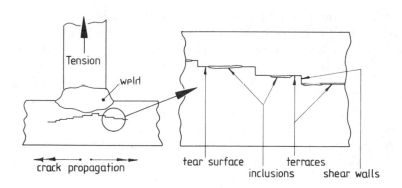

Length of inclusions (nominal dimensions)
large .... 50μm or longer.
medium .... 8 – 50μm.
small ....8μm or shorter.

Fig. 1.13 Sketch of a lamellar tear.

Fig. 1.14 Lamellar tearing. (Courtesy of The Welding Institute.)

(a) Factors affecting lamellar tearing

The factors governing lamellar tearing can be listed as:

(1) 'Quality' of the steel, i.e. quantity and shape of inclusions
(2) Rolling temperature of steel
(3) Strain induced in steel due to welding, determined by weld size.

The shape of inclusions in the steel is very important, and it appears that round inclusions are far less harmful than the long sharp-ended variety. These inclusions are often residues of sulphides, oxides and silicates formed in the molten steel by an additive made to reduce oxygen content and to refine grain structure. The shape of inclusions can be controlled by additions such as rare earths, Ca, Zr or Ti. 'Z' quality steels can now be specified to have a maximum sulphur content of 0.02% having in some cases only half that value in the final product. Z quality steel is further defined in (c) below.

According to the WRC bulletin 31 on lamellar tearing, the thermal expansion properties of steel during mill rolling also play a part in the problem. Non-metallic inclusions tend to plasticize and then liquefy at 900°C and above. The selection of rolling temperature will decide whether they are present in planar form. Research into the behaviour of inclusions suggests that microstresses at the boundary of inclusions are non-existent in the case of manganese sulphide (due to a favourable thermal coefficient of expansion) but aluminate inclusions give rise to high stresses. There is also some evidence to imply cracking around inclusions and fine cracks connecting inclusions.

When the steel plate is welded it is affected in two ways; the steel acquires a heat-affected zone in the region of the weld, and strain accumulates in a small area adjacent to the weld, due to weld shrinkage. When plates at 90° are welded together the location of the weld is important (see comments on connection design). The weld should be positioned so that the strain is shared by the two plates. The connection should be specially designed to resist service loads without the occurrence of lamellar tearing and may be constructed as described in the following section.

(b) Connection design for lamellar tearing

Joints with low ductility are more susceptible to lamellar tearing, therefore the less stiff and restrained a connection, the less LT is likely to occur. Ductility can be 'built in' to a connection by good design and intelligent use of weld. The tensile strength of weld and steel plate should match closely, but unfortunately this

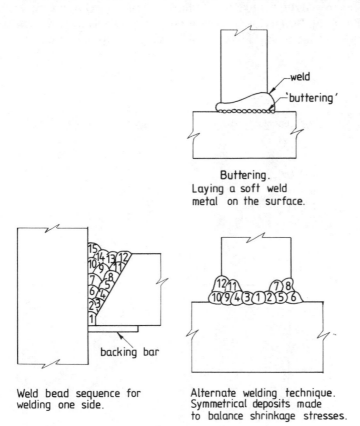

Buttering.
Laying a soft weld
metal on the surface.

Weld bead sequence for
welding one side.

Alternate welding technique.
Symmetrical deposits made
to balance shrinkage stresses.

Fig. 1.15

generally means that the yield point of the weld metal is higher. This means that the base metal yields to stress before the weld metal, thus contributing to LT. Large welds will shrink more than small welds, and this also adds to the problem. There is also cumulative shrinkage in the length of a weld. Good joint design to prevent LT should take into account all the above points and attempt to introduce ductility and low weld shrinkage strains. Basically one should design the smallest weld sizes with the lowest yield strength possible and, where larger welds must be used, employ good welding sequences such as buttering and proper stringer bead sequence (see Fig. 1.15). In addition, preheat and maintenance of interpass temperatures are helpful for fatigue and brittle fracture considerations, and also minimize LT. When considering beam/column welded connections, try to avoid welding the beam web to the column flange; instead weld beam flanges, and use bolts in the beam web. This avoids a concentration of strain in the column flange. Highly restrained connections should be avoided. Some typical methods of welding to avoid LT are sketched in Fig. 1.16.

One way to avoid LT is to add a T shape or cruciform forging or casting to the susceptible joint. Butt welds are then used between the forging/casting and plate. Butt welds not involving through-thickness strain are not prone to LT.

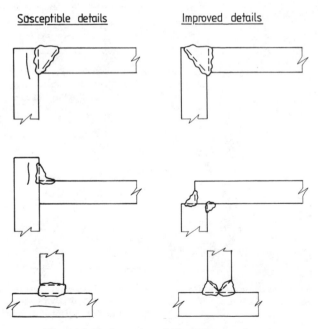

Fig. 1.16 Design to avoid lamellar tearing.

(c) Checks and testing of plate before and after fabrication

In order to lessen expected LT problems, 'Z' steel could be specified. Nos 1, 2, 3, below, typify the features of a Z steel.

Consult a metallurgist in order to:

(1) Specify 'clean' plate (sulphur content not greater than 0.02%)
(2) Specify chemical additions and rolling temperature to obtain small inclusion populations and round inclusion shapes (generally too expensive for small quantities of steel)
(3) Take samples for thru' thickness tensile tests (not expensive); specify a minimum reduction in test area of 25 to 30%.

1.4 NON-DESTRUCTIVE TESTING

1.4.1 General

Various methods of testing steelwork for defects are available as indicated below. In general these methods are employed on the 'heavier' type of steel construction where extensive welding is carried out. The tests are applied by qualified inspectors working to a recognized welding code [40].

1.4.2 Penetrant testing

For structural purposes, two types of penetrant are suitable, water washable, and solvent removable. The member to be tested should be as smooth as possible and free from dirt or grease. Occasionally the member may require to be 'dressed' so that the test result can be more conclusive. The surface is brushed with penetrant dye and shortly after washed with either water or solvent as applicable.

The penetrant is drawn into surface breaking defects such as cracks and drawn out by painting an absorbent developer on to the surface, thus showing the location of flaws and magnifying their appearance. Fluorescent dye methods are used for greater sensitivity.

1.4.3 Magnetic particle testing

This type of test depends on the magnetizing property of ferrous materials. A dry powder of magnetic particles (or liquid with magnetic particles in suspension) is poured or sprayed on to a clean test area. The material is then magnetized locally, usually by applying an electro magnet to the work, and the location of surface defects is rendered apparent by accumulated particles collecting in the line of cracks due to the local distribution of the magnetic flux. The magnetizing technique must be correctly chosen to match the type of defect being sought.

1.4.4 X and gamma radiography

X and Gamma rays have very short wavelengths and thus have the property of passing quite readily through dense substances. A radiograph can distinguish features of lower density from those of higher density, and can be used to obtain photographic images of welds and plate to check for their soundness and freedom from non-metallic inclusions, or defects such as pores or cracks.

The energy source and a special photographic film are placed either side of the item to be checked. The image obtained of a defect may show as a variation in the darkening or a series of blackish dots on a grey background. The rays pass through foreign matter more readily than steel, causing the photographic film to blacken more at the locations of impurities. X rays are produced by the bombardment of a 'target' in a vacuum with free electrons. The emission of X rays depends on the atomic number of the target (usually tungsten). With higher applied voltage, the wavelength of X rays shortens, increasing their penetrating power. Gamma rays are produced by the radioactive decay of isotopes where intensity is related to the half life of the isotope used, and cannot be varied as X rays can. Gamma rays usually have shorter wavelength than X rays and have greater penetrating power. Two commonly used isotopes are Iridium 192 (used for radiographing steel from 12 to 60 mm in thickness), and Caesium 137 (used for radiographing steel from 25 to 100 mm in thickness). The sensitivity of the radiograph can be gauged by the use of wire indicators called 'image quality indicators' IQI, placed on the area to be radiographed. The sensitivity is the ratio of the smallest wire that appears clearly, divided by the plate thickness, multiplied by 100% (see BS3971). Safety precautions that must be observed are laid down in *Ionising Radiations Regulations* (*sealed sources*), 1969, published by HMSO, and the *Code of Practice for Site Radiography* (by a Working Group under the chairmanship of HM Factory Inspectorate), published by Kluwer–Harap Handbooks, London 1976.

1.4.5 Ultrasonic sound

Ultrasonic equipment uses sound well above the audible range. The sound is transmitted from a 'probe' which employs a specially cut crystal of a material such as quartz, to which a suitable voltage is applied, causing the crystal to vibrate. Good contact between the probe and the steel is ensured with a coating of grease or liquid. The steel surface must not only be clean and dry, but also smooth, to ensure efficient operation. The probe is generally made to act as a receiver as well as transmitter. The signal received back from the steel is usually monitored on a cathode ray tube. Compression sound waves are generated in the steel by placing the probe at 90° to the steel surface, and shear waves are generated by placing a wedge shaped perspex shoe between probe and steel. These sound waves are reflected and refracted in a way that is similar to the behaviour of light in a transparent solid. Welds are often inspected by means of reflecting an 'inclined' signal off the far face of the steel plate in order that the signal meets the weld at approximately 90° and returns. On major steelwork, plate is scanned for defects local to the weld before welding, and the welds are also scanned later.

Ultrasound equipment is portable, easy to use, and can detect planar defects that cannot be detected with radiography. The equipment can usually be used only by a skilled operator, as it takes training and experience to recognize the significance of the displayed information.

Modern developments, such as the 'P scan' system, allow a visual representation of the defects to be assessed.

1.4.6 Acoustic emission

Strains in metals caused by the manufacturing process, fabrication, in service deformation or propagation of cracks, cause energy to be released, some of which is manifest as noise in the solid.

Transducers used to monitor this behaviour are usually adjusted to detect waves in the bandwidth 100–300 kHz. The study of emissions due to crack growth in steel is of particular interest. In general the formation of a plastic zone at a crack tip is accompanied by emission, which increases as the crack (plastic zone) enlarges. There is a close relationship between plastic strain and acoustic emission for metals in that emission occurs only when the maximum previous stress level experienced by the specimen is exceeded. This is known as the 'Kaiser' effect. The emission information received by the equipment is processed into an intelligible form, commonly by a method called ringdown counting, which involves sorting the emission signal into batches of progressively larger order (for instance, multiples of 10). In many steels it has been found that there is a shift towards higher emission amplitudes as failure approaches. This feature is utilized in monitoring crack growth on existing structures where it is important that the structure remain in service (e.g. power stations) and service conditions are such that crack growth is virtually bound to occur. Acoustic emission techniques have also been used to detect inclusions in submerged arc welds during production, and may be more commonly used for general fabrication in the near future. The application of acoustic emission techniques is, however, still limited to specialist consultants, and awaits the availability of equipment that is relatively easy to use and that supplies results that do not require a highly trained operative. Techniques in this field are in a state of rapid advancement.

1.5 STRESS RELIEF

1.5.1 Heat treatment

Normalized steels as produced by the mills are usually 'tougher' than 'as rolled' steels, and are to be preferred where brittle fracture is a possibility. Normalizing is the process whereby steel is heated uniformly to approximately 900° C, removing residual stresses due to rolling; it also causes recrystallization, hence refining the grain structure of the steel. If the use of normalized steel is not possible, then there are two basic options open to the engineer. First, a fracture mechanics assessment can be made of the defects found in the structure on completion. If these results are not acceptable, the effect of reducing residual stresses to a low value should be considered. BS5500:1985 gives guidance; the structure may be considered as a long pressure vessel. Should these calculations be acceptable, then part or all of the structure concerned could be 'stress relieved' by heating, whereby the steel is heated uniformly to 600° C to reduce residual stress to a maximum value of approximately 25% yield. It should be noted that thermal stress relief of structures that have already been welded is not normally attempted, but it is possible and does improve resistance to brittle fracture (but not the result of a Charpy test). Care should be taken that heating of parts of a connected structure does not introduce unacceptable strains elsewhere. No further welding *whatsoever* can be allowed on a normalized or stress relieved structure unless further post-welding stress relief is intended.

1.5.2 Vibratory stress relief

The vibratory stress relieving (VSR) process involves the use of high force vibratory equipment to induce one or more resonant states into a metal fabrication. The nominal applied strains in the process are elastic but at points in the member where stress concentration exists, the process causes local overload and plastic flow on a very minute scale takes place. The total effect is to reduce all internal stresses. Size, stiffness, and weight are factors that determine the natural frequency of a member of 'rigid' assembly. Modern VSR equipment is based on an AC design, with a delivery force of not less than 900 kg at 50 Hz and above, with a minimum frequency range of 0 to 200 Hz. The use of the equipment involves attaching the vibrator to the individual steel member with heavy duty clamps, and the vibrator is then 'scanned' slowly from 0 to 200 Hz for approximately ten minutes. At points when resonance takes place, vibration is allowed to continue for about 2000 cycles. A second scan is usually carried out to further reduce internal stresses. Small rigid structures have two or three resonant frequencies, and long flexible structures may have as many as ten resonant frequencies.

The VSR technique offers a considerable advantage in that it reduces internal stresses without altering the metallurgical condition of the component. The equipment is relatively cheap. Further information on the subject can be found in reference 38.

REFERENCES

Selection of paint system

1. Building Research Establishment (1973) *Painting: Iron & Steel*. Digest No. 70, BRE, Garston.
2. Constrado (1980) *Protection of Structural Steelwork from Atmospheric Corrosion*, 2nd Edn. Constrado, Croydon.
3. BS5493 (1977) *Code of Practice for Protective Coating of Iron and Steel Structures against Corrosion*. British Standards Institution, London.

Selection of a suitable steel

4. BS4360 (1986) *Specification for Weldable Structural Steels*. BSI, London.
5. ISO/R630 (1980) *Structural Steels*. International Standards Organisation, ISO, Geneva.
6. Euronorm 25 (1972) *General Structural Steels*, Commission of European Community, Brussels.
7. ASTM Standards for steels (individual steel standards for specific steels, too numerous to note here). ASTM, Philadelphia.

History of brittle fracture

8. Tipper, C. F. (1962) *The Brittle Fracture Story*. Cambridge University Press.
9. Richards, K. G. (1971) *Brittle Fracture of Welded Structures*. Welding Institute.
10. *Engineering Methods for the Design and Selection of Materials against Fracture*. Wessel, E. T., Clark, W. G. and Wilson, W. K. (1966). Final Technical Report for the US Army. Westinghouse Research Laboratories.

Fracture mechanics testing

11. BS5447 (1977) *Methods of Test for Plane Strain Fracture Toughness (KIc) of Metallic Materials*. BSI, London.
12. BS5762 (1979) *Methods for Crack Tip Opening Displacement (CTOD) Testing*. BSI, London.
13. BS131:Part 2 (1972) *The Charpy V-notch Impact Test on Metals*. BSI, London.

Application of fracture mechanics methods

14. PD6493 (1980) *Guidance on Some Methods for the Derivation of Acceptance Levels for Defects in Fusion Welded Joints*. BSI, London.
15. Harrison, R. P., Loosemore, K. and Milne, I. (1976) *Assessment of the Integrity of Structures Containing Defects*. The Central Electricity Generating Board.

Fatigue

16. Richards, K. G. (1969) *Fatigue Strength of Welded Structures*. Welding Institute.

17. Munse, W. H. (1964) *Fatigue of Welded Steel Structures* (Ed. La Motte Grover). Welding Research Council, New York.
18. Influence of details on fatigue of structures (1968) *JASCE*, pp. 2679–97.
19. Gurney, T. R. (1976) Fatigue design rules for welded steel joints. *Weld. Inst. Res. Bull.* **17**, May.
20. BS5400 (1980) *Steel, Concrete and Composite Bridges: Part 10: Code of Practice for Fatigue*. BSI, London.
21. Fisher, J. W. and Struik, H. J. A. (1974) *Guide to Design Criteria for Bolted and Riveted Joints*. Wiley, New York.

Stress corrosion and corrosion fatigue

22. *Stress Corrosion Cracking and Corrosion Fatigue in Offshore Structures*. Hockenhull B. S. (1976). Event No. 86, Joint Offshore Conference. European Federation of Corrosion.
23. Scott, P. M. and Silvester, D. R. V. (1975) *The Influence of Seawater on Fatigue Crack Propagation Rates in Structural Steel*. UK Offshore Steels research project. Dept. of Energy.

Steel properties and fracture mechanics

24. Cottrell, A. H. (1964) *The Mechanical Properties of Matter*. Wiley, New York.
25. Knott, J. F. (1973) *Fundamentals of Fracture Mechanics*. Butterworths, London.

Welding distortion

26. *Control of Distortion in Welded Fabrications* (1968). Welding Institute.
27. Blodgett, G. W. (1980) *Solutions to Design of Weldments*. The James F. Lincoln Arc Welding Foundation, Ohio.

Lamellar Tearing

28. Pope, C. W. and Card, K. J. (1976) Detection of LT by Ultrasonic Testing *Non-Destructive Testing, Australia* **13**, Nov.–Dec.
29. Anon. (1974) More insight to lamellar tearing. *Welding Design Fabricator*, **41**, January.
30. Anon. (1973) Lamellar tearing of welded connections. *Civ. Engr. (NY)*, **43**, December.
31. Jubb, J. E. M. (1971) Lamellar tearing. Bull. 168, Welding Research Council, New York.
32. Thornton, C. H. (1973) Quality control in design and supervision can eliminate LT. *Eng. J. AISC*, 4th quarter.

Non-destructive testing

33. BS2600 (1973) *Radiographic Examination of Fusion Welded Butt Joints in Steel*. Over 50mm up to and including 200mm Thick. BSI, London.
34. BS3923 (1972) *Methods for Ultrasonic Examination of Welds*. Automatic Examination of Fusion Welded Butt Joints in Ferritic Steels. BSI, London.

35. BS6072 (1981) *Method for Magnetic Particle Flaw Detection.* BSI, London.
36. BS6443 (1984) *Method for Penetrant Flaw Detection.* BSI, London.
37. Various authors (1974) *Acoustic Emission.* IPC Science and Technology Press, Guildford.

Stress relief

38. Claxton, R. A. (1979) Vibratory stress relieving – practice and theory. *Proc. Conf. on Heat Treatment – Methods and Media.* Institution of Metallurgists, pp. 34–45.

Welding Inspection and Testing

39. BS709 (1983) *Methods of Destructive Testing Fusion Welded Joints and Weld Metal in Steel.* BSI, London.
40. ANSI/AWS D1.1 – 81. Structural Welding Code (American National Standard).

The Authors wish to acknowledge the help received by the following engineers and metallurgists in reviewing and commenting on this chapter: John Barker, Bryan Blanchard, Chris Blow, Ken John, John Newman, Bill Streeten.

Structural elements

2.1 INTRODUCTION

The elements of a structure are designed to carry their loads and/or moments while ensuring safety and economy in the use of steel. Economy in design is not necessarily attained by a member sized to obtain a minimum weight of steel. Simple fabrication details using a heavier section often lead to greater overall economy.

The most important criteria for steel design are those concerned with strength, stability and deflection. The *ultimate* limit state defines the stress at which a member will fail by yielding (strength requirement), whereas the geometric properties of the member may be such that buckling would occur at a lower stress value (stability requirement). Deflection limits (serviceability limit state) are usually based on the maximum amount of deflection that can be accepted visually, although excessive deflection may, for instance, cause damage to finishes or even alter the structural shape (and hence the loading pattern). More unusual loading situations are discussed in Chapter 1.

The design method usually entails determining 'factored' loading and 'collapse' moments, which are then used to determine member sizes. The members are checked for deflection using the elastic theory and unfactored actual loads (see Table 2.1).

The behaviour of steel under load can best be appreciated by studying the steel stress/strain graph (Fig. 2.1a). Within the elastic range of steel, loading or moments on a steel member will not cause permanent deformation. Within the plastic range, permanent deformation takes place (i.e. strain continues with no increase in stress). These features are illustrated in Fig. 2.1b for a moment carrying beam.

The cross-sectional properties of members are classified in the code under the heading 'plastic', 'compact', 'semi-compact' and 'slender'. 'Plastic' denotes a section with proportions that allow it to develop a *plastic hinge and sufficient rotation* to allow redistribution of bending moments in the structure. 'Compact' sections can develop *plastic moment* but local buckling takes place before rotation can occur. 'Semi-compact' sections have proportions that allow yield at the extreme fibres only, without buckling. 'Slender' sections reach some value less than yield at the extreme fibres, dependent upon buckling considerations. For plastic design, only 'plastic' sections should be used, so that plastic action can be attained in the structure as a whole, and the entire cross section yields

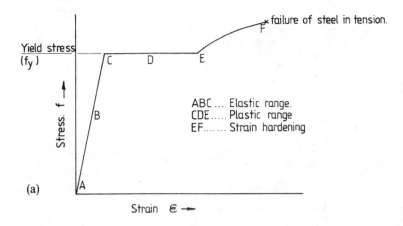

ABC .... Elastic range.
CDE ..... Plastic range
EF .... Strain hardening

(a)

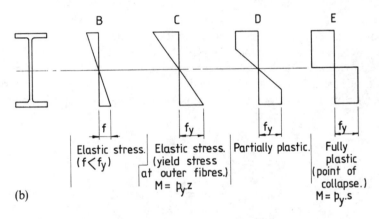

Elastic stress.
(f<f$_y$)

Elastic stress.
(yield stress
at outer fibres.)
M= p$_y$.z

Partially plastic.

Fully
plastic
(point of
collapse.)
M= p$_y$.s

(b)

Fig. 2.1 (a) Tensile stress/strain graph for steel. (b) Stress diagrams for bending moment on a steel beam.

(where $M_c = p_y S$ but not greater than $1.2p_y Z$) (see Fig. 2.2) (M_c = moment capacity; p_y = design strength; S = plastic modulus; Z = elastic modulus.)

The limiting proportions of sections, and yield stress of the steel, govern their shape classification; they are given in Table 7, Clause 3.5.2 of the code.

Residual stresses occur in all steel members owing to differential cooling after hot rolling, or to local heating caused by welding followed by cooling. Typical distribution of these stresses is indicated in Fig. 2.3. The presence of these stresses is undesirable because in the case of welded components they cause the steel to yield at a low value.

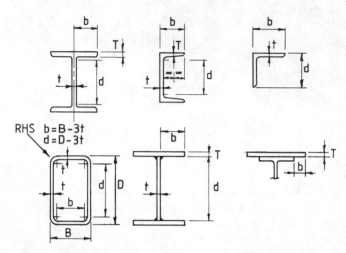

Fig. 2.2 A selection of b/T proportions for sections (from Fig. 3 in the code) for use in determining their classification – plastic, compact or semi-compact.

In brief, the design of structural elements must be considered in the context of the whole structural scheme – a typical procedure being structural analysis, followed by element design, followed by design of connections. If the connection cannot meet the design criteria it may necessitate a change in the structural scheme, or, more probably, the selection of a different structural element.

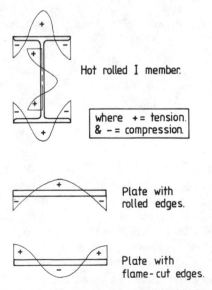

Fig. 2.3 Typical residual stress distributions in the longitudinal direction of rolling.

2.2 BEAMS

Beams are the elements of a structure that generally span between columns and are designed to resist bending/torsional moments and shear forces. This section outlines the main factors to be taken into account in the design of beams although many of the considerations apply equally to column design.

Where beams are subjected to bending and axial loads, the design procedure is described in Section 2.3.4.

2.2.1 Bending

One type of beam failure occurs when its moment carrying capacity or plastic moment is reached. The moment carrying capacity can only be utilized provided that failure does not occur at a lower load level by either local or overall buckling. Therefore a check should be carried out to ensure that the *buckling resistance moment* (M_b) exceeds the *moment capacity* (M_c).

The bending moment at which a beam fails by buckling when subjected to a uniform moment is called the 'elastic critical moment' (M_E).

An 'equivalent' uniform moment ($\bar{m}$) can be calculated by multiplying the maximum applied bending moment at the end of a beam by a factor m, whose value is stated in Table 13 of the code (for I, H and channel shapes). The *buckling resistance moment* M_b represents the maximum value of equivalent uniform moment that is permitted for a beam of given length, being not larger than the product of the plastic modulus S and the bending strength p_b. The bending strength of a member is dependent on the equivalent slenderness λ_{LT} (λ_{LT} is a function of the slenderness ratio $\lambda = L_E/r_y$). The effective length L_E of a beam is governed by the restraint conditions at the ends of the member.

Procedure for determining the bending strength P_b for a beam
(examples given in Section 2.2.8)

(1) Define beam end restraint condition and select a value of L_E, based on code Tables 9 and 10. Note p_y for steel grade.
(2) Calculate $\lambda = L_E/r_y$.
(3) Select values for u and n based on code clauses 4.3.7.5. and 4.3.7.6.
(4) Calculate v as follows: obtain λ/x, where $x = D/T$ for I, H or channel (clause 4.3.7.7 in the code).
(5) Calculate $\lambda_{LT} = n.u.v.\lambda$. Obtain value of v from code Table 14.
(6) Obtain value of p_b from code Table 11 or 12 as appropriate (for λ_{LT} against p_y).

The *moment capacity* of a section (M_{cx} or M_{cy}) is dependent on the beam shape. For plastic and compact sections, plastic stress distribution may be used, and the moment capacity $M_c = p_y S$ but not greater than $1.2p_y Z$. For semi-compact

Table 2.1 DEFLECTION: ref. Code Table 5
$M = $ max. bending moment due to imposed load only (unfactored) in kNm.
$L = $ span in metres.
Deflection limits are defined in the Code. Using these limits and the formula $\delta = kML^2/I$, for a simply supported beam, a list of factors for minimum I is given in this table.

Beam type	Loading	Min. I value cm^4 (gross)	
		for $\delta = \dfrac{L}{360}$	for $\delta = \dfrac{L}{200}$
Uniform cross section	central point load.	14.5 $M.L$	8.1 $M.L$
Simply supported	U.D.L.	18.2 $M.L$	10.1 $M.L$
	Uniform moment for any type of loading.	21.8 $M.L$	12.1 $M.L$
Fixed ended	end point load	29.1 $M.L$ $\Big\}$ for $\delta = \dfrac{L}{180}$	
cantilevers	U.D.L.	10.9 $M.L$	

Beam overhanging
one support.

dimensions in metres

for $\delta = \dfrac{L}{180}$

$14.3\dfrac{P.a}{L}(2aL + 2a^2)$

Crane gantry girder	central point load.	vertical deflection due to static wheels for $\delta = L/600$
Simply supported.		$I = 32.67\,M.L$

Final size of girder to be checked using Code formula.

horizontal deflection due to horiz. wheel loads.
$I_{y.y}$ top flange only
for $\delta = L/500$
$I = 20\,M.L$

Column type	Loading	Min. I value cm^4 (gross)
Fixed both ends	horiz. point load at top.	for $\delta = H/300$ $I = 23.8\,M.H$
Fixed at top, bottom pinned	horiz. point load at top.	for $\delta = H/300$ $I = 47.6\,M.H$
Pinned both ends i.e. gable cols.	horizontal U.D.L.	for $\delta = H/200$ (Authors suggest this as a limit.) $I = 10\,M.H$

sections, elastic stress distribution is used, and $M_c = p_y Z$. Slender sections (i.e. narrow flanges, deep webs) are analysed on the basis of a lower design strength, p_y, the bending stress diagram being assumed by the designer and the section checked accordingly. A typical check would include an assumption that moment is carried by the flanges only, and that the web carries the shear. Slender sections are generally unsuitable for locations of high bending moment, and designers generally avoid their use, as the lower design strengths tend to produce higher steel weights and less economy. When thin flanges are used, local buckling is allowed for by using lower design strengths as indicated in clause 3.6 of the code.

2.2.2 Lateral torsional buckling of beams

The lateral torsional buckling of a beam is the tendency of the beam to twist sideways as shown in Fig. 2.4. If a beam is subjected to an increasing uniform moment, at a critical value of moment M_E it will collapse suddenly by buckling. The beam's tendency to behave in this way depends upon the following factors:

(1) Beam sectional properties
(2) End fixing of beam
(3) Beam span
(4) Imposed moment
(5) Distances between lateral restraints on the compression flange of the beam
(6) Direction and eccentricity of applied load relative to the shear centre.

If buckling occurs while the beam is still elastic, the value M_E can be evaluated as follows:

$$M_E = \frac{M_p \pi^2 E}{\lambda_{LT}^2 p_y} \text{ (code Appendix B.2.2)}$$

(for a symmetrical I section simply supported in the lateral plane and loaded by a continuous moment) where λ_{LT} is the equivalent slenderness of the member (code 4.3.7.5).

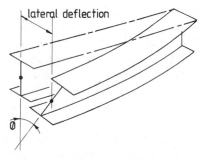

lateral deflection

Fig. 2.4 Lateral torsional buckling.

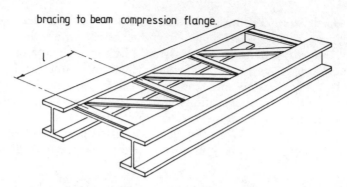

Fig. 2.5 Lateral bracing to beams.

Destabilizing loads on the compression flange of beams (for instance a horizontal force applied to the compression flange at mid-span) create a measure of instability to the beam, causing the section to buckle at a lower value of applied moment than would otherwise be the case. The code (Table 9) makes an allowance for this by allocating longer effective lengths where destabilizing conditions exist. A convenient method of defining a member's resistance to buckling failure may be given by the ratio $\sqrt{M_p/M_E}$, where M_p is the full plastic moment (i.e. $M_p = S_x p_y$, where $S_x =$ plastic modulus about $x - x$ axis and $p_y =$ design strength of steel). Experiments show that *stocky* beams with a $\sqrt{M_p/M_E}$ value less than 0.4 can develop M_p, whereas *slender* beams with an $\sqrt{M_p/M_E}$ value greater than 1.2 fail at moment values near M_E.

If lateral bracing is introduced to a previously unbraced beam, the λ_{LT} value is substantially reduced. Lateral bracing is taken to a point near (or at) the compression flange of the member (see Fig. 2.5). Such bracing is usually designed to resist a lateral force of 1% of the force in the compression flange provided, that the stiffness of the entire bracing system is such that it limits lateral deflection to less than 1/25 of the lateral deflection that would otherwise occur.

2.2.3 Equivalent uniform moment

Beams with fixed ends, and parts of beams between lateral restraints should be checked for equivalent uniform moment. The moment diagram for the portion of member considered is compared with code Table 18. The diagram from the table is as given in Fig. 2.6. The larger moment at either end of the member (or sub-member) is then factored down by an appropriate value m in the table, giving the equivalent uniform moment $\bar{M}$. The value of $\bar{M}$ should not exceed the value of M_b (buckling resistance moment described in previous section). The values of m quoted in the table apply to section shapes with equal flanges only. In the case of unsymmetrical section shapes and cantilevers without lateral restraint, $\bar{M}$ is taken as the maximum moment. A member bent in *single curvature*

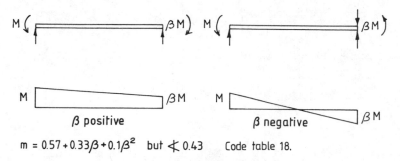

$$m = 0.57 + 0.33\beta + 0.1\beta^2 \quad \text{but} \not< 0.43 \quad \text{Code table 18.}$$

Fig. 2.6 Equivalent uniform moment factor m.

to a uniform moment is more prone to buckling than a similar member bent in *contraflexure* with similar values of end moment, and the value of $\bar{M}$ compared to M_b provides a check against buckling.

2.2.4 Shear (code 4.2.3)

The allowable 'plastic' moment on a beam section is reduced by a high shear force. Usually it is assumed that the central portion of the web (on an I beam) carries the shear, the remainder of the section carrying the moment. Providing the average shear stress F_v on the shear area ($=tD$ for rolled sections) does not exceed $0.6P_v$, ($0.36p_y$), no reduction in the moment capacity is necessary. However, if (in exceptional circumstances) F_v is greater than $0.6P_v$, the reduced moment capacity is given for plastic and compact sections (code 4.2.6) by:

$$M_c = (S - S_v\rho_1)p_y \quad \text{but} \not> 1.2p_yZ$$

where S_v = plastic modulus of web
$\quad\quad\ = tD^2/4$ for a rolled I section
$\rho_1 = (2.5F_v/P_v) - 1.5$

and t, D and S are web thickness, overall depth and plastic modulus of the section respectively.

Reduction in moment capacity does not apply for semi-compact and slender sections, as they are designed using elastic theory.

Shear in a section is usually assumed to be distributed in an elastic fashion in accordance with the formula (based on the theory of simple bending):

$$f_q = \frac{VA\bar{y}}{It_w} \quad \text{(See Fig. 2.7)}$$

where f_q = maximum elastic shear stress
$\quad\quad V$ = shear force

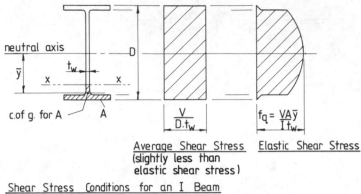

neutral axis

$\bar{y}$

t_w

c.of g. for A   A

$\dfrac{V}{D.t_w}$

$f_q = \dfrac{VA\bar{y}}{I\,t_w}$

<u>Average Shear Stress</u>   <u>Elastic Shear Stress</u>
(slightly less than
elastic shear stress)

<u>Shear Stress Conditions for an I Beam</u>
acting both normal to, & longitudinal to, the beam.

Fig. 2.7 Distribution of shear in a beam.

t_w = total thickness of section at neutral axis (i.e. for beam,
   web thickness; for tube $2 \times$ wall thickness)
A = area between plane xx and extreme fibre
$\bar{y}$ = distance between neutral axis and centroid of area A
I = second moment of area of the section.

However, it is usually sufficient to check average shear stress for the majority of I sections, the difference between maximum elastic shear stress and average shear stress being quite small. The code notes an exception in the case of sections whose webs vary in thickness, or contain holes for services (code 4.2.4) in which case f_q must not exceed $0.7p_y$. Loading on structural sections should attempt to meet the condition whereby the line of action of the applied load passes through the shear centre. The shear centre is located such that if an applied load passes through it, no twisting of the section occurs. This is best illustrated by considering the effect of loading on a shape whose shear centre is eccentric, such as a channel (see Fig. 2.8). The location of the shear centre for open sections with one axis of symmetry may be determined by obtaining the I_x values for portions of the structural shape and the distance (x) of the centroid of individual shapes from an assumed line. Thus the location of the shear centre e of the whole (measured at $90°$ to the yy axis), is a distance measured from the line, where $e = -\Sigma I_x x/\Sigma I_x$. The location of the shear centre for any other sections can be found by equating the torsional moment caused by unit loading to the product of the unit load and the unknown distance e.

2.2.5 Web buckling

Web buckling may occur at locations of high point loads, either under a point load in the span, or at the supports. It is assumed that the load from a seating

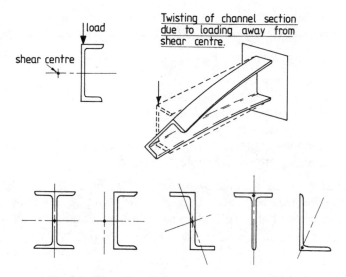

Fig. 2.8 Shear centres for a selection of sections.

Web buckling
Code 4.5.2.1

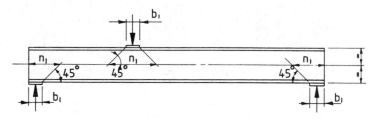

Fig. 2.9 Web buckling.

spreads at an angle of 45°, being a near approximation to the situation that occurs in practice. This is illustrated in Fig. 2.9. Buckling resistance

$$P_W = (b_1 + n_1)tp_c \qquad \text{(code 4.5.2.1)}$$

where b_1 = stiff bearing length

n_1 = length obtained by dispersion at 45° through half the depth of the section.

t = web thickness

p_c = compressive strength from code 4.7.5 treating the web as a column with $a = 5.5$ (code Table 27c)

The slenderness of the unstiffened web may be taken as $2.5\,d/t$ only if a beam frames into the member under consideration with the top flange and web connected either by bolting or welding. Otherwise the radius of gyration must be taken as $t/\sqrt{12}$ and the value of λ taken as L/r. (i.e. $d\sqrt{12}/t$).

2.2.6 Bearing capacity

The local capacity of a web in bearing is calculated on a load dispersion through a beam flange of 1 to 2.5 (see Fig. 2.10). However, since *web buckling* is a more common design problem than crushing in a bearing, the check on bearing is often limited to verifying the load-carrying capacity of web stiffeners, added primarily as a protection against web buckling. However, if the bearing capacity of an unstiffened web is exceeded stiffeners must be provided, designed to the requirements of code 4.5.3.

2.2.7 Lattice girders

Lattice girders are suitable for large spans, preferably loaded at node points to avoid local bending in the top or bottom chord members. Their most common use is in roof construction designed to carry the roof dead and imposed loads, or in some cases to carry wind only. The depth of the girder is governed by the type of loading. For light loads a depth of approximately one-tenth of the span

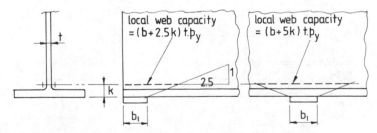

Fig. 2.10 Bearing – dispersion of load through a flange.

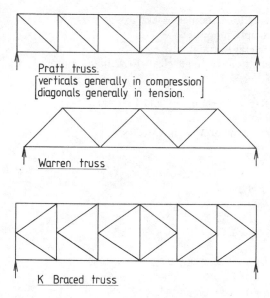

Pratt truss.
[verticals generally in compression]
[diagonals generally in tension.]

Warren truss

K Braced truss

Fig. 2.11 Types of lattice girder.

may be suitable; for heavy loads the depth may need to be one-sixth of the span. The general rules for design are:

(1) Keep the compression members short;
(2) Diagonals should not be at too shallow an angle as this increases the gusset size (if used).

If gussets are not employed in the design, difficulties could arise at the fabrication stage where members would have to be cut to significantly finer tolerances.

Smaller lattice girders often employ symmetrical members for the top and bottom chords but may use single angles or even bars for the internal members. Larger lattice girders are provided with symmetrical members throughout, sometimes using 'I' beams for the chords. The bracing system selected is often a matter of the designer's preference (some typical types are shown in Fig. 2.11). Sometimes additional members are added to the bracing in order to carry tension when load reversals occur, or to reduce the unbraced lengths of members. A design example of a lattice girder is given in Section 2.2.8.

2.2.8 Beam design examples

Example 2.1

Design a suitable beam for the following:
Factored total moment = 20 kNm

Unfactored 'Imposed load' moment = 8kNm
Span = 6m = i.e. simply supported
No intermediate lateral restraint
'Framed' end connection (i.e. angle cleats to web and bottom beam flange)

SOLUTION
Use grade 50 steel ($p_y = 355\,\text{N/mm}^2$), and I section.
Nominal effective length = $1.0\,L$ (code Table 9)
Equivalent uniform moment $\bar{M} = \text{max.} = 20\,\text{kNm}$

Try 203 × 133UB25

$\lambda = 6000/31.0 = 193.5 \quad D/T = 26.0 = x$
$n = 1.0$ (conservatively) (code 4.3.7.6)
$u = 0.9$ (code 4.3.7.5)
Find v: $N = I_c/(I_c + I_t) = 0.5$
$\qquad \lambda/x = 193.5/26 = 7.44$
$\qquad v = 0.72$ (code Table 14)
$\lambda_{\text{LT}} = nuv\lambda = 1 \times 0.9 \times 0.72 \times 193.5 = 125.4$ (code 4.3.7.5)
$p_b = 96.5\,\text{N/mm}^2$ (code Table 11)
Check buckling resistance moment M_b
$M_b = S_x p_b$
$\qquad = 259.8 \times 10^3 \times 96.5/10^6\,\text{kNm}$
$\qquad = 25.1\,\text{kNm} > 20\,\text{kNm}$ satisfactory.
Check deflection
Minimum I value to restrict deflection to $L/360$
(due to imposed loading)
$= 18.2\,\text{ml.}$ (Refer Table 2.1)
$= 18.2 \times 8 \times 6 = 873.6\,\text{cm}^4 < 2356$ provided.
An alternative method is to calculate the deflection from the standard formula, and ensure that it does not exceed $L/360$.
Use 203 × 133UB25, grade 50

Example 2.2

Design a beam for the following criteria:

Span = 6m = L_E. No lateral restraint in span; framed connection at ends; simply supported; grade 43 steel.
UDL–Imposed = 10.67 kN/m of span (unfactored)
$\qquad\qquad$ Dead = 1.08 kN/m of span (unfactored)

SOLUTION (factors from Table 1.0 or code Table 2)
Factored imposed load = $WL\gamma_f = 10.67 \times 6 \times 1.6 = 102.0\,\text{kN}$
$\quad$ Factored dead load = $WL\gamma_f = 1.08 \times 6 \times 1.4 = 9.1\,\text{kN}$

$\qquad\qquad\qquad$ Total load on beam = 111.1 kN

$M_{max} = WL/8 = 111.1 \times 6/8 = 83.3\,\text{kNm}$

Unfactored imposed load $M_{max} = WL^2/8 = 10.67 \times 6^2/8 = 48\,\text{kNm}$

$L_E = 1.0\,L$ $M_B\,(P133) = 65 < 83.$ (code Table 9)

Try $305 \times 165\text{UB}40$ $p_y = 275\,\text{N/mm}^2$ (code Table 6)

$\lambda = 6000/38.5 = 156,$ $x = D/T = 29.8$

$\left. \begin{array}{l} n = 1.0 \\ u = 0.9 \end{array} \right\}$ (conservatively)   (code 4.3.7.5)

Find v: $N = 0.5$ for I section

$\quad \lambda/x = 156/29.8 = 5.23$

$\quad v = 0.81$ (code Table 14)

$\lambda_{LT} = nuv\lambda = 1 \times 0.9 \times 0.81 \times 156 = 114$ (code 4.3.7.5)

$p_b = 103.4\,\text{N/mm}^2$ (interpolating between values) $\rightarrow M_B = 64.6$ (code Table 11)

min. $S_x = m/p_b = 83.3 \times 10^3/103.4 = 805.6\,\text{cm}^3$

For a $305 \times 165\text{UB}40$, $S_x = 624.5 < 805.6$; try larger section.

Try $406 \times 178\text{UB}54$ $p_y = 275\,\text{N/mm}^2$ $M_B = 104$ (code Table 6)

$\lambda = 6000/38.5 = 155,$ $x = D/T = 37$ 38.5

$n = 1.0$ (conservatively)

$u = 0.9$ for I section. $.872$ (code 4.3.7.5)

Find v: $N = 0.5$ for I shape

$\quad \lambda/x = 155/37 = 4.19$ 4.03

$\quad v = 0.85$ 0.86 116.2 (code Table 14)

$\lambda_{LT} = nuv\lambda = 1 \times 0.9 \times 0.85 \times 155 = 105.4$

$p_b = 116.4\,\text{N/mm}^2$ (interpolating)   (code Table 11)

Min. $S_x = 83.3 \times 10^3/116.4 = 716\,\text{cm}^3 < 1048$ (provided) $\rightarrow \therefore M_b = 105.4\,(\text{for } n=1)$

i.e. $M_b < \bar{M}$ (this condition is satisfied)

Min. $I = 18.2 \times 48 \times 6 = 5241\,\text{cm}^4 < 18626$ (provided)

Check effects of shear

Shear force at support = Total load/2

$111.1/2 = 55.5\,\text{kN}$

$A_v = tD = 7.6 \times 402.6 = 3060\,\text{mm}^2$ (code 4.2.3)

Average shear stress

$f_y = 55.5 \times 10^3/3060 = 18.14\,\text{N/mm}^2.$

Allowable shear stress $= 0.36\,p_y$

$\qquad\qquad\qquad = 0.36 \times 275 = 99\,\text{N/mm}^2$

$\qquad\qquad\qquad > 18.14\,\text{N/mm}^2$

Section is adequate for average shear stress.

Use $406 \times 178\text{UB}54$

Example 2.3 (*Lattice girder*)

Design the constituent members for the lattice girder (assume self weight of members to be relatively minor and neglect in calculation) shown in Fig. 2.12, using grade 43 steel and grade 8.8 bolts.

SOLUTION – see empirical design rules (code 4.10)

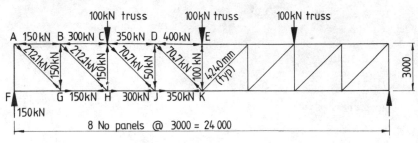

Fig. 2.12 Pratt truss (factored loads shown).

Design of diagonals AG and BH:

Let $p_y = 275 \text{ N/mm}^2$ (code Table 6)

Max. axial load in diagonals = 212.1 kN tension

Nett area required = $212.1 \times 10^3/275 = 771.3 \text{ mm}^2 = 7.71 \text{ cm}^2$

Try 2/65 × 50 × 8 angles with M16 bolts (18 diameter

holes, long leg bolted)

For one angle,

$A_e = a_1 + a_2 [3a_1/(3a_1 + a_2)]$

a_1 = sectional area of connected leg   (code 4.6.3.1)

a_2 = sectional area of unconnected leg

$a_1 = (65 - 18 - 4)8 = 344 \text{ mm}^2$ (code 4.6.3.2. Note.)

$a_2 = (50 - 4)8 = 368 \text{ mm}^2$

$\therefore A_e = 3.44 + 3.68 [3 \times 3.44/(3 \times 3.44 + 3.68)] = 6.15 \text{ cm}^2$

A_e for 2 angles = $2 \times 6.15 = 13.3 \text{ cm}^2 > 7.71 \text{ cm}^2$

Centres of connections from ends of member   (code 4.6.3.3)

= 9 × smallest leg length of angles

= 9 × 50 = 450 mm.

Also use one connection at mid-length.

Transverse force at connections = 1% axial load   (code 4.6.4)

= 0.01 × 212.1 = 2.1 kN

which can be seen to be less than the shear capacity of an M16 bolt (i.e. 58.9 kN – see Fig. 3.1 in Chapter 3)

Use 2/65 × 50 × 8 angles long leg bolted with M16 bolts,

separated with 8 mm packs 40 mm square 450 mm from each

end and at mid-length

Design of diagonals CJ and DK:

Max. axial load in diagonals = 70.7 kN tension

Nett area required = $70.7 \times 10^3/275 = 257.1 \text{ mm}^2 = 2.57 \text{ cm}^2$

Use 2/65 × 50 × 8 angles – as above

Design of lower chord:
Let $p_y = 275 \text{ N/mm}^2$ (thickness $\ngtr 16$ mm)   (code Table 6)
Max. axial load in chord JK $= 350$ kN tension
Nett area required $= 350 \times 10^3/275 = 1273 \text{ mm}^2 = 12.73 \text{ cm}^2$
Try $2/100 \times 65 \times 8$ angles long leg bolted with M16 bolts

(18 diameter holes)
For one angle, $a_1 = (100 - 18 - 4)8 = 624 \text{ mm}^2$, $a_2 = (100 - 4)8 = 768 \text{ mm}^2$ (code 4.6.3.2)
$A_e = 6.24 + 7.68 \, [3 \times 6.24/(3 \times 6.24 + 7.68)]$
   $= 11.69 \text{ cm}^2$ for one angle
A_e for 2 angles $= 2 \times 11.69 = 23.38 \text{ cm}^2 > 12.73 \text{ cm}^2$
Centres of connections from ends of members $= 9 \times 65 = 585$ mm.
Use $2/100 \times 65 \times 8$ angles long leg bolted with M16 bolts, separated with 8 mm packs

40 mm square 585 mm from each end and at mid-length.

Design of verticals BG and CH:
Max. axial load in member $= 150$ kN compression.
Assume effective length $= 0.85L$ (code Table 24)
Try $2/75 \times 50 \times 8$ Ls connected at intervals (code 4.7.13) along their length.

$L = 0.85 \times 3000 = 2550$ mm.
Area of angles $= 2 \times 9.41 \text{ cm}^2 = 18.82 \text{ cm}^2$.
Slenderness $\lambda = \sqrt{(\lambda_m^2 + \lambda_c^2)}$ but not less than $1.4\lambda_c$ (code 4.7.9(c))
$\lambda_m = L_E/r_y$ compound member $= 2550/21.9 = 116.4$
Max. distance between bolt centres $= 50 \times 8 = 400$ mm $=$ sub. L (code 4.7.9(d))
$\lambda_c = $ sub. L/r_{vv} for one angle ... not greater than 50.
   $= 400/10.7 = 37.4 < 50$
Strut divided into 8 equal bays of 375 mm.
$1.4\lambda_c = 1.4 \times 37.4 = 52.3$
   $\lambda = \sqrt{(116.4^2 + 37.4^2)} = 122.3 > 52.3$
   $p_y = 275 \text{ N/mm}^2$
$\therefore p_c = 94.6 \text{ N/mm}^2$ and allowable load on   (code Table 27(c))
strut $= 1882 \times 94.6/10^3 = 177.9$ kN > 150
Use $2/75 \times 50 \times 8$ Ls with 8 mm packing plates at intervals of 375 mm

Intermediate connections
Longitudinal shear per interconnection $= 0.25 \, Q\lambda c$ (code 4.7.13.1(f))
where $Q = 0.01 \times$ axial load in member
    $= 0.01 \times 150$
    $= 1.5$ kN
Longitudinal shear per connection $= 0.25 \times 1.5 \times 37.4 = 14$ kN   (code 4.7.13.1(g))
Try M16 grade 8.8 bolts

(16 mm diameter bolts are minimum size for back to back struts)   (code 4.7.13.1(f))
From Table 3.1.1 in Chapter 3, shear capacity of one bolt
   $= 58.9$ kN > 14 kN

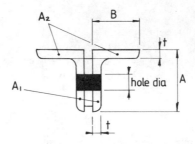

Fig. 2.13 Double angle tie member.

Check bearing capacity
Bearing capacity of bolt $= dtp_{bb} = 16 \times 8 \times 970/10^3$
$\qquad\qquad\qquad = 124.2 > 14\,\text{kN}$ (code 6.3.3.2 and Table 32)
Bearing capacity (ply) $= dtp_{bs} = 16 \times 8 \times 460/10^3$
$\qquad\qquad\qquad = 58.9 > 14\,\text{kN}$ (code 6.3.3.3 and Table 33)
OR $= \frac{1}{2}etp_{bs}$ $\qquad = 0.5 \times 22.5 \times 8 \times 460/10^3 = 36 > 14\,\text{kN}$
$(e = 1.40d = 22.5\,\text{mm})$
Use 1 No. M16 grade 8.8 bolt with 45×45 pack $\times 8$ thick at 400 mm centres

Design of verticals DJ and EK:
Working to the same procedure as outlined for verticals BG and CH, the following sizes are recommended: $2/65 \times 50 \times 8$Ls with 8 mm packs, 45 mm square at 400 mm centres with 1 No. M16 grade 8.8 bolt at each interconnection.

Design of upper chord:
Maximum axial load $= 400\,\text{kN}$ compression (DE). Upper chord could be designed as two channels battened together on top and bottom flanges. However, design upper chord as universal column member. Truss and roofing to provide restraint via truss shoes to top member.
L_x (panel points) $= 0.85 \times 3000 = 2550\,\text{mm}$
L_y (between truss shoes) $= 0.85 \times 6000 = 5100$ mm
Try 152×152UC30
$\lambda_{xx} = 2550/38.2 = 66.75$
$\lambda_{yy} = 5100/67.5 = 75.56$. This governs.
Axial stress in member $= 400 \times 10^3/38.2 \times 10^2 = 10.47\,\text{N/mm}^2$
For a rolled H-section buckling about the yy axis, use Table 27c (code Table 25)
Then compressure strength $= 170\,\text{N/mm}^2 > 104.7$ (code Table 27c)
Use 152×152UC30 for top chord member.

The foregoing example is by way of illustration only, and the engineer should always bear in mind the availability of structural section types, and the best structural use for each type. In this example it may be cheaper to use tube bracing. In the final design the effect of bending stresses due to member self weight would be considered. Deflection due to dead load on the truss should be calculated and the truss precambered upwards to counteract the deflected amount.

2.3 COMPRESSION MEMBERS

2.3.1 Development of column theory

The buckling of long columns was originally investigated by Euler (1759), the 'buckling' or Euler stress being given as

$$p_e = \frac{\pi^2 E}{(L/r_y)^2}$$

(Fig. 2.14)

which can also be written as:

$$p_e = \frac{\pi^2 EI}{L^2 A}$$

where

p_e is the Euler stress
E is the modulus of elasticity
A is the cross-sectional area of the column

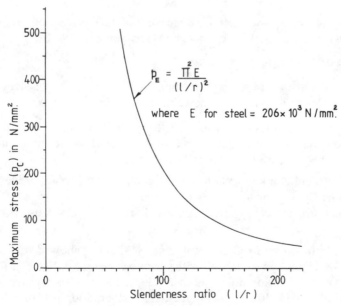

Fig. 2.14 The Euler curve (for steel columns).

L is the length of the column
r_y is the radius of gyration about the yy axis
I is the moment of inertia about the yy axis.

The Euler load of a column is the load which induces initial buckling. Euler's formula did not take account of the yield stress of individual steels, or the practical imperfections of columns in service. However, it formed the mathematical basis from which present day column design graphs are derived. The column treatment used in Britain for the 50 years prior to BS5950 has been based on the work of Ayrton and Perry (1886). Their assumptions can be listed as follows.

(1) Column assumed to be pin-ended;
(2) Initial buckling of single sinusoidal curvature;
(3) Column behaves elastically;
(4) No residual stress;
(5) Collapse occurs when the stress on the extreme fibre of the concave side reaches yield point.

Predicted failure of the column is given by the Perry formula:

$$(p_y - p_c)(p_e - p_c) = \eta p_e p_c$$

where p_y = yield stress (design strength)
   p_c = mean stress at failure
   p_e = Euler stress as noted above.

η is a non-dimensional factor and is a measure of the linear amount of deviation from straightness. If a tolerance of $L/1000$ is assumed

$$\eta = 0.001 \ (c/r)(L/r)$$

where c/r = a function of the section shape for a given axis (xx or yy)
   L/r = slenderness ratio of the column.

The value of η provides the means of fitting the theory to the experimental results. However, strut strengths predicted with the above formula and η value will tend to be inaccurate for the following reasons:

(1) The first yield is taken as failure – (column could carry more load);
(2) Strain hardening for low L/r is neglected;
(3) Residual stresses are ignored. (Residual stresses could cause a premature failure.)

Tests undertaken by Professor Robertson (1925) [6] showed that a value of $\eta = 0.003 \ L/r$ compared quite well with the experimental data (i.e. a Robertson constant c/r of 3). This finding was then adopted as a basis for strut design in BS449 and BS153. Godfrey (1962) suggested a further modification to the Perry–Robertson curve [7] and in the interests of economy the value of η in BS449 was changed to

$$\eta = 0.3 \left(\frac{L}{100r} \right)^2$$

Current column design curves are based on an extensive column research programme and theoretical work carried out at Graz by Beer and Schultz (1970); and at Cambridge by Young (1971). In the code design curves are provided for four different types of column section. The difference in behaviour of various column shapes is so great that it has been found essential to produce separate curves. The original European curves were defined by polynomial expressions to fit experimental curves. However, in the code the curves are represented by

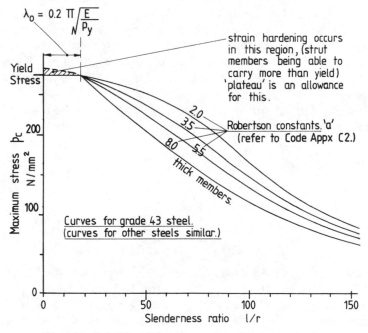

Modified Perry formula is:–
$(p_y - p_c)(p_E - p_c) = \eta \cdot p_E \cdot p_c$

which can be written:

$p_c = \dfrac{p_E \cdot p_y}{\emptyset + \sqrt{\emptyset^2 - p_E \cdot p_y}} \geqslant p_y$, where $\emptyset = \dfrac{p_y + (\eta + 1)p_E}{2}$

& $\eta = 0.001a(\lambda - \lambda_o) \not< 0$

plateau, limiting slenderness ratio

$\lambda_o = 0.2 \, \pi \sqrt{\dfrac{E}{p_y}}$

strain hardening occurs in this region, (strut members being able to carry more than yield) 'plateau' is an allowance for this.

Yield Stress

2.0
3.5   Robertson constants. 'a'
8.0   5.5   (refer to Code Appx C2.)

thick members.

Maximum stress p_c N/mm²

200

100

Curves for grade 43 steel.
(curves for other steels similar.)

0

0   50   100   150

Slenderness ratio   l/r

Fig. 2.15 Column axial stress against slenderness ratio.

the modified Perry formula which is more convenient. The new value for η thus became:

$$\eta = 0.001a \, (\lambda - \lambda_0) \ldots \text{but not less than zero}$$

where

η is the Perry factor
a is the Robertson constant (value 2.0, 3.5, 5.5, or 8.0 depending on section type and axis considered)
λ is the slenderness L/r
λ_0 is the limiting slenderness, taken as $0.2\sqrt{(\pi^2 E/p_y)}$.

The present design curves allow for imperfections such as column straightness and profile, and the existence of residual stresses. It can be seen from the curves that a plateau exists at low L/r values to allow for the effect of strain hardening on short columns (see Fig. 2.15).

2.3.2 Column effective lengths (L_E)

The compressive strength of a column is dependent on the slenderness ratio (L_E/r), where L_E = effective length and r = radius of gyration. For simple one storey structures Table 2.2 illustrates common situations of end restraint for columns. The type of end restraint determines the effective length, so that a column pinned at each end has an effective length of the entire length of the column, whereas a column 'fixed' in position and direction at each end has an effective length of $0.7 \times$ total length. The effective length of a column is thus that portion of the column that lies between points of contraflexure or 'pin' connections.

There are two broadly different types of structure from the point of view of determining column effective lengths, which are sway frames and non-sway frames.

The amount of fixity provided by 'fixed' moment carrying beam to column connections is governed by the relative I (second moment of area) values of the beams and column. For multi-storey construction a joint restraint value k for both ends of the column may be calculated where

$$k_{1,2} = \frac{\text{Total column stiffness at joint}}{\text{Total stiffness of all members at joint}}$$

The effective length of the column considered can be found by reference to Figs 23 and 24 in Appendix E of the code where the former figure represents L_E values for rigidly braced frames and the latter, for unbraced frames.

Guidance on effective lengths for typical column types and end restraints in single storey buildings is given in Appendix D of the code.

Table 2.2 Nominal effective lengths for struts based on code Table 24.

Column Types

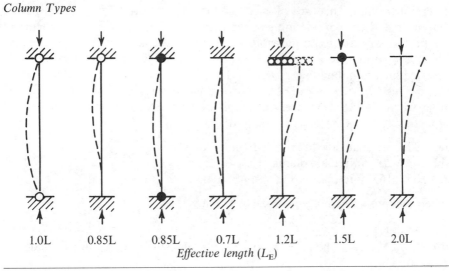

1.0L	0.85L	0.85L	0.7L	1.2L	1.5L	2.0L

Effective length (L_E)

Conditions of restraint at column ends

Effectively held in position
but not restrained in direction. (pin)

Effectively held in position
and partially restrained in direction.

Effectively held in position
and restrained in direction. (fixed)

Not held in position
and effectively restrained in direction.

Not held in position
and partially restrained in direction.

Not held in position
or restrained in direction. (free)

2.3.3 Axial load in compression

The cross-sectional shape of a column influences its behaviour under axial compression. The tendency of any shape to buckle under axial compression is related to the c/r value of the shape (c/r equals distance from centroid to extreme edge of section divided by the radius of gyration). Values of c and r are measured at 90° to each other.

With the application of axial load a lateral deflection occurs which induces a moment in the column; the moment equals the product of axial load and the lateral deflection. This type of effect, named 'second-order' effect, is of greater importance when considering slender members. Generally, the most efficient shape to carry axial compression is the tube, but as connection of I beam to tube may involve a special connection, the universal column is normally selected for use in buildings.

To illustrate the above, let us consider three steel column shapes of approximately the same cross-sectional area, all having an effective length of 4 metres, (grade 43 steel at 275 N/mm^2):

(a) 193.7 mm diameter tube with 10 mm wall thickness
(b) 203 × 203 universal column (UC) 46 kg/m
(c) 305 × 165 universal beam (UB) 46 kg/m

Their compressive strengths or capacities are:

(a) 1366 kN⎫
(b) 967 kN⎬ using the appropriate code Table 27 determined from Table 25.
(c) 800 kN⎭

For axially loaded members of trusses and space frames, tubes are often selected for bracing, and the chords, if the moments are low. Angle, tee and channel shapes are relatively inefficient as columns, but are often used for bracing due to ease of connection.

The design and methods of connections involving circular tube, rectangular hollow section (RHS) and I beam chord/column members are presented in tabular form in Chapter 3, Tables 3.4 to 3.7 inclusive.

Procedure for obtaining allowable axial load on a column
(*examples given in Section 2.3.6*).

(1) Obtain length of member and end fixing type
(2) Determine effective length based on code Table 24 (Table 2.2 in this book) (for multi-storey building see code Appendix E)
(3) Select a trial section and steel grade. Note p_y from code Table 6
(4) Note r_y and r_x
(5) Calculate $\lambda_{xx} = L_x/r_x$ and $\lambda_{yy} = L_y/r_y$ (note the higher value of λ)
   Note: $\lambda_{xx} \ngtr 180$ ⎛unless bracing $\ngtr 350$ for cross bracing,⎞
         $\lambda_{yy} \ngtr 180$ ⎝resisting wind, then $\ngtr 250$ for K bracing. ⎠
(6) Determine Robertson constant for the steel section from code Table 25 (i.e. type a, b, c, or d column action)
(7) Using higher value of λ, and value of p_y, obtain compressive strength p_c, from code Table 27 (a, b, c, or d)
(8) Maximum axial load = $p_c A$ kN (in absence of moment and shear)

(9) For case including moment, shear and axial load or any combination of these, see code on combined stresses, and below

(10) See typical examples in Section 2.3.6.

2.3.4 Combined bending and axial load

Axial load on a section reduces the allowable moment capacity. For 'plastic' and 'compact' sections approximate reduced values of plastic modulus 'S' may be determined from the formulae quoted in published tables, such as the BCSA/Constrado handbook or as follows:

For rolled I and H sections:
where $n = F/Ap_y$

where $n \leqslant 0.2$ $\quad S_{rx} = (1 - 2.5n^2) S_x$

$\quad\quad n \geqslant 0.2$ $\quad S_{rx} = 1.125 (1 - n) S_x$

$\quad\quad n \leqslant 0.447$ $\quad S_{ry} = (1 - 0.5n^2) S_y$

$\quad\quad n \geqslant 0.447$ $\quad S_{ry} = 1.125 (1 - n^2) S_y$

where $S_{rx} =$ reduced plastic modulus in x axis
$\quad\quad S_{ry} =$ reduced plastic modulus in y axis
$\quad\quad F =$ applied load.

Thus the reduced moment capacity $M_r = S_r p_y$ and moment capacity $M_c = S p_y$. Interaction formulae covering the load–moment possibilities for strength are given in the code with the proviso that the combined sum of ratios of axial load and moment in each axis should not exceed 1. Thus an increase in any one of the variables (axial load, moment about xx, moment about yy) can be seen to reduce the limit for the other value(s). For plastic and compact members in particular (see code 4.2.5) the value of M_c is limited to the smaller of $p_y S$ or $1.2 p_y Z$; this is to ensure that the minor axis moment is kept to a value below that which would permit yield stress to occur at service loading.

For members subject to combined moment and axial load a local capacity check must be made at points of maximum bending moment and axial load (usually at the column ends) as follows:

$$\left(\frac{M_x}{M_{rx}}\right)^{z_1} + \left(\frac{M_y}{M_{ry}}\right)^{z_2} \not> 1 \text{ for plastic and compact sections} \quad \text{(code 4.8.3.2)}$$

where M_{rx}, M_{ry} are reduced moment capacities in presence of axial load, and z_1 and z_2 are constants (2.0 and 1.0 respectively for H and I sections).

For a simplified approach for compact sections; and for semi-compact, and slender sections, local capacity should meet the following:

$$\frac{F}{A_g p_y} + \frac{M_x}{M_{cx}} + \frac{M_y}{M_{cy}} \not> 1$$

where M_{cx}, M_{cy} are moment capacities in the absence of axial load, and A_g is the gross cross-sectional area (nomenclature otherwise as noted previously).

Overall buckling

Plastic and compact sections with very low slenderness ratio are able to carry moment up to plastic action, where moment capacity equals the product of plastic modulus and steel yield stress. Members with higher slenderness ratio in either axis are subject to buckling consideration. For instance, I and H members with high moment about the major (xx) axis will collapse by deflection of the flanges, in a twisting motion about the section centre. Collapse due to moment in the minor axis is not accompanied by twisting, but results in a localized deformation of the flanges.

Overall buckling may be checked by either a simplified method or by a somewhat lengthier approach. The simplified method involves less calculation, with conservative results, but the member should be checked at locations of maximum deflection due to bending – i.e. at mid-point for a strut in single curvature, or at quarter points for a strut in double curvature.

In the simplified approach the following relationship should be satisfied:

$$\frac{F}{A_g p_c} + \frac{mM_x}{M_b} + \frac{mM_y}{p_y Z_y} \not> 1$$

where A_g = gross cross-sectional area
p_c = compressive strength
M_b = buckling resistance moment (see Section 2.2.1)

The simplified approach indicates the type of buckling behaviour that may be anticipated with I or H sections, where the limits of axial load and major axis moment decrease with increase in the L/r_y of the member, but the limit of minor axis moment remains constant at the maximum of elastic action on the shape. The above formula also indicates that major axis buckling is dependent on the lateral torsional resistance of the column flanges. The same considerations apply to the more complex formula, although greater accuracy can be achieved in some cases, leading to economy.

The more exact approach for checking overall buckling requires that:

$$\frac{mM_x}{M_{ax}} + \frac{mM_y}{M_{ay}} \not> 1 \qquad \text{(code 4.8.3.3.2)}$$

where m = equivalent uniform moment factor obtained from code Table 18 (see also Section 2.2.3)
M_{ax} = maximum buckling moment about the xx⎫ axis in the presence of axial load.
M_{ay} = maximum buckling moment about the yy⎬ axis in the presence of axial load.⎭

$M_{ax} = M_{cx}(1 - F/P_{cx})/(1 + 0.5F/P_{cx}) \not> M_b(1 - F/P_{cy})$
$M_{ay} = M_{cy}(1 - F/P_{cy})/(1 + 0.5F/P_{cy})$

$M_b =$ buckling resistance moment, value dependent on the lateral torsional resistance of the member.

$P_{cx} =$ member capacity for major axis buckling (using λ_x)

$P_{cy} =$ member capacity for minor axis buckling (using λ_y)

2.3.5 Built-up columns

When the area required for a single rolled column section is in excess of that available, then cover plates may be added, thus increasing the area, moment of inertia and radius of gyration of the section.

This method can be utilized when the same core section of a column is required for several storey heights in a building. The column 'core' listed in the BCSA Steel Handbook on sections is, for instance, the most appropriate core section to use if a heavier section than the heaviest rolled column section is required.

When flange plates are added to a rolled I section, the plate overhang should be limited to the proportions indicated in the code, 3.5.5; Table 7 and the force on the weld joining the plate to the flange should be checked by finding the elastic shear stress at that point (illustrated in Fig. 2.7).

Tall buildings with a free interior utilized as a working area (such as workshops for heavy equipment) are often designed with 'compound' columns comprising two or three rolled sections laced together. Usually single rolled sections are not adequate in such locations. Lacing or battening is intended to provide a smaller slenderness ratio for the column, but not smaller than the value for the xx axis. If

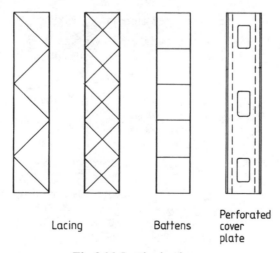

Lacing   Battens   Perforated cover plate

Fig. 2.16 Latticed columns.

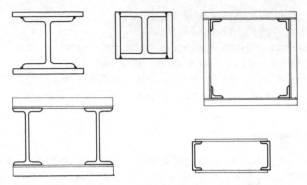

Fig. 2.17 A selection of built-up column shapes.

columns are latticed on four sides (as for instance square, angle frame towers) very large loads can be carried by relatively small steel members. This feature is often utilized for transmission towers, and water tank supports (see Figs 2.16 and 2.17).

2.3.6 Column design examples

Example 2.4 (Fig. 2.18)

Design a suitable column in grade 50 steel.
Consider upper length of column with larger moments.
Try 203 × 203UC46

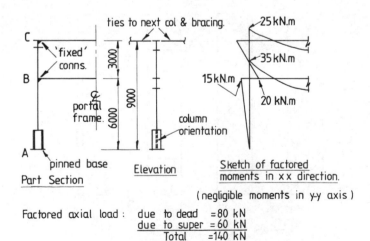

Fig. 2.18 Column design, Example 2.4.

$p_y = 355$ N/mm^2 (thickness of steel less than 16 mm).   (code Table 6)
Section properties (from BCSA/Constrado Handbook):
$A = 58.8$ cm^2
Plastic modulus $S_y = 230$ cm^3; $S_x = 497.4$ cm^3
Elastic modulus $Z_y = 151.5$ cm^3; $Z_x = 449.2$ cm^3
$r_y = 51.1$ mm; $r_x = 88.1$ mm
Section classification:
Check the proportion of the cross-section by ready reference tables
or as below: (code 3.5.2, Table 7 and Fig. 3)
$\varepsilon = (275/355)^{\frac{1}{2}} = 0.88$
$b/T = 203.2/2 \times 11 \times 0.88 = 10.5 > 9.5 < 15$
∴ outstand element of compression flange is semi-compact.
$d/t\varepsilon = 160.8/7.3 \times 0.88 = 25 < 79$
∴ web, with neutral axis at mid-depth is plastic.
The section classification is thus semi-compact.
Slenderness ratio, λ and compressive strength, p_c
$\lambda = L/r$
$\lambda_{yy} = 9000/51.1 = 175 < 200$ (worst case)
$\lambda_{xx} = 2.0 \times 6000/88.1 = 136$
$\dfrac{F}{A} = \dfrac{140 \times 10^3}{58.8 \times 10^2} = 23.8$ N/mm^2 (at base)
Assume this stress occurs at point B.
The Robertson constant a determines which strut table should be used
This has been interpreted in the code so that for this rolled H
section, less than 40 mm thick, for axis xx Table 27(b) is the one to
use and for axis yy, Table 27(c) is appropriate.   (code Appendix C and Table 25)
For $\lambda_{xx} = 136$, and $p_y = 355$, $p_{cx} = 92.8$ N/mm^2 (interpolating)   (code Table 27(b))
For $\lambda_{yy} = 175$, and $p_y = 355$, $p_{cy} = 55$ N/mm^2 (code Table 27(c))
Check local capacity:
$F/A_y p_y + M_x/M_{cx} \leqslant 1.0$; Moment capacity $M_{cx} = p_y S \ngtr 1.2 p_y Z$ (code 14.8.3.2)
$M_{cx} = 355 \times 497.4 \times 10^{-3} = 176.5$ kNm ($< 1.2 \times 449.2 \times 355$)   (code 4.2.5)
∴ $F/A_g p_y + M_x/M_{cx} = 23.8/355 + 25/176.5 = 0.21 < 1.0$
Check overall buckling:
$\dfrac{mM_x}{M_{ax}} + \dfrac{mM_y}{M_{ay}} \ngtr 1$ (code 4.8.3.3.2)
where M_{ax} is the maximum buckling moment about the major axis in the presence of
axial load. Note: No bending in yy axis.
Determine value of v (slenderness factor):
$N = I_{cf}/I_{cf} + I_{tf} = 0.5$ (code 4.3.7.5)
$\lambda/x = 175(D/T) = 175/18.5 = 9.46$
$\quad v = 0.65$ (code Table 14)
$\quad n = 1.0$ (Member not loaded in length)   (code Table 13)
$\lambda_{LT} = nuv\lambda_{yy}$ (code 4.3.7.5)
$\quad\quad = 1.0 \times 0.9 \times 0.65 \times 175$
$\quad\quad = 102.4$
Determine p_b ($p_y = 355$ N/mm^2):
$p_b = 134.2$ N/mm^2 (interpolating)   (code Table 11)

Determine $\bar{M}_x$ (equivalent uniform moment about xx axis);
$m = 1.0$ (conservative assumption)   (code Tables 13 and 18)
$\bar{M}_x = 1.0 \times 25$ kNm $= 25$ kNm
Buckling resistance M_b:
$M_b = S_x p_b = 497.4 \times 10^3 \times 134.2/10^6 = 66.75$ kNm   (code 4.3.7.3)
$M_{ax} =$ lesser of:

$$M_{cx}(1 - F/P_{cx})/(1 + 0.5F/P_{cx}) \tag{1}$$

or

$$M_b(1 - F/P_{cy}) \tag{2}$$

From (1) 176.5 $(1 - 140 \times 10^3/92.8 \times 58.8 \times 10^2)/(1 + 0.5 \times 140 \times 10^3/92.8 \times 58.8 \times 10^2)$
$= 116.3$ kNm
From (2) 66.75 $(1 - 140 \times 10^3/59 \times 58.8 \times 10^2) = 39.8$ kNm

$$\frac{mM_x}{M_{ax}} + \frac{mM_y}{M_{ay}} \not> 1$$

$$\frac{25.0}{39.8} + 0 = 0.63 < 1.0$$

∴ section is adequate against overall buckling
Use 203×203UC46 grade 50

Example 2.5 (Fig. 2.19)

$M_{xx} = 10 \times 7 = 10$ kNm, $M_{yy} = 0$
Axial load $= 135$ kN. Max. plate thickness $= 15$ mm.
Grade 43 steel, $p_y = 275$ N/mm².

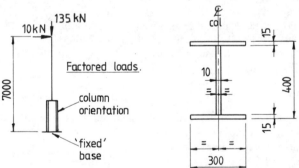

Note:
Negligible loads &
moments in y-y direction.

Section thro' welded plate column.
(gr. 43)

Check that the above column is adequate for the applied loads &
moments. Neglect deflection.

Fig. 2.19 Column design, Example 2.5.

Section properties:

$I_y = 2 \times 1.5 \times 30^3/12 + 37 \times 1^3/12 = 6753\ \text{cm}^4.$

$A = 30 \times 1.5 \times 2 + 37 \times 1 = 127\ \text{cm}^2.$

$I_x = (30 \times 40^3 - 29 \times 37^3)/12 = 37589\ \text{cm}^4$

$r_x = \sqrt{(I_x/A)} = \sqrt{(37589/127)} = 17.2\ \text{cm}$

$r_y = \sqrt{(I_y/A)} = \sqrt{(6753/127)} = 7.3\ \text{cm}$

$z_x = I/y = 37589/20 = 1879\ \text{cm}^3$

$s_x = bd^2/4 + BT(d + T) = 0.25 \times 1 \times 37^2 + 30 \times 1.5(37 + 1.5) = 2075\ \text{cm}^3$

Slenderness ratios:

$\left.\begin{array}{l} \lambda_{xx} = 2 \times 7000/172 = 81 \\ \lambda_{yy} = 2 \times 7000/73 = 192 \end{array}\right\}$ Effective length $L_E = 2L$ (from Table 2.2)

$F/A = 135 \times 10^3/127 \times 10^2 = 10.6\ \text{N/mm}^2$

For a welded plate I or H section the code strut tables to be used are:

(code Table 25)

Axis xx, Table 27(b); axis yy, Table 27(c)

$p_y = 275 - 20 = 255\ \text{N/mm}^2$

Hence $p_{cx} = 120.5\ \text{N/mm}^2$ and $p_{cy} = 44.2\ \text{N/mm}^2$ (code Tables 27(b) and (c))

$p_{cx} = p_{cx}A = 170.5 \times 127 \times 10^2/10^3 = 2165.4\ \text{kN}$ (code 4.7.5)

$p_{cy} = 44.2 \times 127 \times 10^2/10^3 = 561.3\ \text{kN}$

Check section type

$p_y = 275\ \text{N/mm}^2$

$b/T = 145/15 = 9.67$

$d/t = 370/10 = 37$

$\varepsilon = (275/275)^{\frac{1}{2}} = 1.0$

Since $8.5\varepsilon < b/T < 13\varepsilon$,   (code Table 7)

the section is semi-compact.

Check shear   (code 4.2.3)

Shear at base of column $= 10\ \text{kN}$

$A_v = 37 \times 1 = 37\ \text{cm}^2$

$p_v = 0.6 \times 275 = 165\ \text{N/mm}^2$ and $0.6p_v = 99\ \text{N/mm}^2$

$f_v = \dfrac{\text{shear}}{A_v} = \dfrac{10 \times 10^3}{37 \times 10^2} = 2.7\ \text{N/mm}^2 < 0.6\,p_v$

Moment capacity

$M_{cx} = p_y Z_x = 275 \times 1879 \times 10^3/10^6 = 516.7\ \text{kNm}$ (code 4.2.5)

Note on compressive strength: For welded section p_y

reduced by 20 N/mm² (4.7.5); this does not apply to p_y (code 4.7.5)

in general though, as noted in code.   (code 4.1.3)

Check local capacity at base   (code 4.8.3.2)

$\dfrac{F}{A_g p_y} + \dfrac{M_x}{M_{cx}} + \dfrac{M_y}{M_{cy}} \not> 1$

$\dfrac{135}{3492} + \dfrac{70}{516.7} + 0 = 0.18 < 1.0,$ adequate

Check overall buckling (use simplified approach)

$\dfrac{F}{A_g p_c} + \dfrac{\bar{M}_x}{M_b} + \dfrac{\bar{M}_y}{p_y Z_y} \not> 1$ (code 4.8.3.3.1)

No bending in yy axis, therefore omit third term.

$\bar{M}_x = M_{max} = 70\,\text{kNm}$

$M_b = S_x p_b$.

Determine p_b.

$\lambda_{LT} = nuv\,\lambda$

$\quad \lambda = 192, u = 1.0, n = 1.0$ (code 4.3.7.5)

$\quad N = 0.5, \lambda/(D/T) = 192/26.7 = 7.2$; hence $v = 0.995$ (code Table 14)

$\lambda_{LT} = 1 \times 1 \times 0.995 \times 192 = 191$

$p_b = 43.6\,\text{N/mm}^2$ (code Table 12)

$$\frac{135 \times 10^3}{127 \times 10^2 \times 44.2} + \frac{70 \times 10^6}{2075 \times 10^3 \times 43.6} = 1.01 > 1.0 \text{ acceptable}$$

Section is adequate for the design loads shown

Example 2.6 (Fig. 2.20)

Try 203×203UC46 (grade 50); $p_y = 355\,\text{N/mm}^2$ (code Table 6)

Section properties

$\quad A = 58.8\,\text{cm}^2; \quad r_x = 88.1\,\text{mm} \quad r_y = 51.1\,\text{mm}$

$\quad S_x = 497.4\,\text{cm}^3 \quad S_y = 230.0\,\text{cm}^3$

$\quad Z_x = 449.2\,\text{cm}^3 \quad Z_y = 151.5\,\text{cm}^3$

$\quad \lambda_{yy} = 0.7 \times 6000/51.1 = 82 \big\}$ Values of L_E

$\quad \lambda_{xx} = 0.7 \times 6000/88.1 = 48 \big\}$ from Table 2.2

For this rolled H section the code strut tables to be used are:

Axis xx Table 27(b): axis yy Table 27(c) (code Table 25)

$\quad p_{cx} = 302\,\text{N/mm}^2; \; p_{cy} = 183\,\text{N/mm}^2$ (code Table 27)

$\quad F/A = 145 \times 10^3/58.8 \times 10^2 = 24.7\,\text{N/mm}^2$

Moments include dead, live, & wind. All values include γ_F factors.

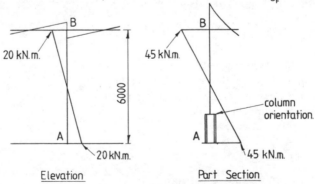

20 kN.m.

6000

A

20 kN.m.

Elevation

45 kN.m.

column orientation.

B

A

45 kN.m.

Part Section

Axial load on column at 'A' = 145 kN.

Design a suitable column in grade 50 steel. (p_y = 355 N/mm²)

Fig. 2.20 Column design, Example 2.6.

$P_{cx} = 302 \times 58.8 \times 10^2/10^3 = 1776\,\text{kN}$
$p_{cy} = 183 \times 58.8 \times 10^2/10^3 = 1076\,\text{kN}$
Check section type
$b = B/2 = 203.2/2 = 101.6;\ b/T = 101.6/11 = 9.2$
$\varepsilon = (275/355)^{\frac{1}{2}} = 0.88$
Limiting proportions for compact section
$b/T \not> 9.5\varepsilon = 9.5 \times 0.88 = 8.36$
and for semi-compact section
$b/T \not> 13 \times 0.88 = 11.4$
Since $8.36 < 9.2 < 11.4$, the section is semi-compact.
Check local capacity at base

$$\frac{F}{A_g p_y} + \frac{M_x}{M_{cx}} + \frac{M_y}{M_{cy}} \leqslant 1 \qquad \text{(code 4.8.3.2)}$$

$$\frac{145 \times 10^3}{58.8 \times 10^2 \times 355} + \frac{45 \times 10^6}{449.2 \times 10^3 \times 355} + \frac{20 \times 10^6}{151.5 \times 10^3 \times 355}$$
$$= 0.77 < 1.0$$

$\therefore$ local capacity of section is adequate
Check overall buckling

$$\frac{\bar{M}_x}{M_{ax}} + \frac{\bar{M}_y}{M_{ay}} \not> 1.0$$

m values from Table 18 of code (or taken conservatively as 1.0)
$\bar{M}_x = 1 \times 45 = 45\,\text{kNm}$
$\bar{M}_y = 1 \times 20 = 20\,\text{kNm}$
and $M_b = S_x p_b$ (code 4.3.7.3)
Determine p_b
$\lambda_{LT} = nuv\lambda.$ (code 4.3.7.5)
  $\lambda = 82, u = 0.9, n = 1.0$
Calculate v
  $N = 0.5$ (code Table 14, Note 1)
$\lambda/x = 82/(D/T) = 82/18.5 = 4.43$
  $v = 0.84$ (code Table 14)
$\lambda_{LT} = 1 \times 0.9 \times 0.84 \times 82 = 62$
  $p_b = 249.8\,\text{N/mm}^2$ (code Table 11)
$\therefore M_b = 497.4 \times 10^3 \times 249.8/10^6 = 124.3\,\text{kNm}$
$M_{ax} = M_{cx}\,(1 - F/P_{cx})/(1 + 0.5F/P_{cx})$
  or $M_b(1 - F/P_{cy})$ (lesser of the two)   (code 4.8.3.3.2)
$M_{ax} = 148.2\,(1 - 145/1776)/(1 + 72.5/1776) = 129.9\,\text{kNm}$
  or $124.3\,(1 - 145/1076) = 107.5\,\text{kNm}$
$M_{ay} = M_{cy}(1 - F/P_{cy})/(1 + 0.5F/P_{cy})$
    $= 50\,(1 - 145/1076)/(1 + 72.5/1076) = 40.5\,\text{kNm}$
$$\frac{\bar{M}_x}{M_{ax}} + \frac{\bar{M}_y}{M_{ay}} = \frac{45}{107.5} + \frac{20}{40.5} = 0.91 < 1.0$$
Use $203 \times 203\text{UC46}$ grade 50

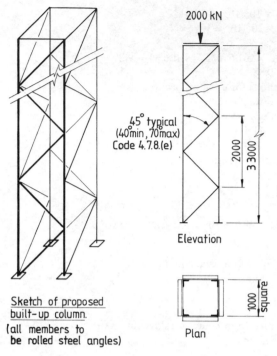

Fig. 2.21 Column design, Example 2.7.

Example 2.7

BUILT-UP COLUMN DESIGN
Design the constituent members of the lattice column shown in Fig. 2.21 in grade 50 steel. Omit design for the reinforced plate at the top for this exercise.

(code 4.7.8)

Consider column legs:
Axial load on built-up column = 2000 kN
Axial load on individual angle legs = 2000/4 = 500 kN
Length L of leg between nodes = 2000 mm
Consider this as fixed ended with $L_E = 0.7L$ (Table 2.2)
Then $L_E = 0.7 \times 2000 = 1400$ mm.
Try $150 \times 150 \times 10$ angle

Area = 2930 mm², $p_y = 355$ N/mm²   (code Table 6)
slenderness ratio $L_E/r_{vv} = 1400/29.7 = 47$
                                    $= \lambda_c < 50$ – satisfactory   (code 4.7.8(g))
Check overall suitability of this angle for the built-up column:
Column L/r, where $r = \sqrt{(I/A)} \simeq b/2 = 1000/2 = 500$
must not be less than $1.4\lambda_c = 1.4 \times 47 = 65.8$ (code 4.7.8(g))
$\lambda_{column} = 33000/500 = 66 > 65.8$

The code strut table for angle sections is Table 27(c)   (code Table 25)
$L/r_{vv} = 47$ ∴ compressive strength, $p_c = 283 \text{ N/mm}^2$ (code Table 27)
Max. allowable load per leg $= 283 \times 2930/10^3 \text{ kN}$
                              $= 829 \text{ kN} > 500 \text{ kN}$

Use $150 \times 150 \times 10$ angle for column legs (grade 50)

Consider lacing members:
Note: bracing in opposite faces matches. No extra allowance for torsional effects necessary.
Axial load on lacing members
$=$ shear force on built-up column + transverse shear due to axial load
    on main member   (code 4.7.8(d))
$= (0 + \sqrt{2}(1\% \text{ of axial load}))/\text{lacings parallel in plane}$ (code 4.7.9(e))
$= \sqrt{2}(0.01 \times 500)/2$
$= 3.5 \text{ kN}$

Try $50 \times 50 \times 8$ angle welded to main member.

Area $= 741 \text{ mm}^2$
Slenderness ratio, $\lambda = 0.85 \, L/r_{vv}$ or $0.7L/r_{aa} + 30$ (code 4.7.10.2 (a), Table 28)
$0.85L/r_{vv} = (0.85 \times 1000\sqrt{2})/9.6 = 125$
$0.7L/r_{aa} + 30 = (0.7 \times 1000\sqrt{2})/14.8 + 30 = 97$
∴ $\lambda = 125 < 180$ so satisfactory
and $p_c = 99 \text{ N/mm}^2$
Max. allowable load on lacing $= 0.8 \times 99 \times 741/10^3 = 58.7 \text{ kN} > 3.5 \text{ kN}$

Use $50 \times 50 \times 8$ angle for lacings and horizontal tie at top of column (grade 50).

2.4 TENSION MEMBERS

Tension members may consist of standard rolled sections, or combinations such as back-to-back angles or channels. It is common to use single angles, channels, tees, or flat plate, especially for bracing. The use of rods is mainly restricted to limiting the vertical deflection of cladding rails by means of 'sag rods'. Eyebars and pin-connected members are used in suspension-type structures, such as suspension bridges and suspended roof systems.

    The slenderness λ of rigid tension members is limited for practical reasons in order to prevent excessive deflection owing to member self-weight or wind gusting. A suitable limit would be $\lambda = L/r_{min} = 250$ for main members, and $\lambda = 350$ for secondary members.

    Difficulties resulting from the possibility of brittle fracture (see Chapter 1) in welded tension elements are mainly avoided by using specified maximum thickness as given in the code Table 4.

2.4.1 Allowable axial load in tension

The design of a member loaded only in tension is limited to a simple check of the axial stress which should not exceed the stated design strength of the steel.

Full account should also be taken of moments induced in the member by bolting connections eccentric to the member axis, and for 'minor' members this is normally achieved by considering a smaller effective area in tension. For the method of computing effective area refer to design example 2.8.

For simple tension only the tension capacity of a member is given by $P_t = A_e p_y$ (code 4.6.1), where A_e is the effective area of the section, either gross area, or a value due to the subtraction of bolt holes, and stress redistribution (code 3.3.3 or 4.6.3) and p_y is the design strength of the steel.

2.4.2 Combined bending and tensile loads (code 4.8)

Bending in two planes with tension in the member axis is generally resisted by a 'plastic' or 'compact' section. It is less likely that a 'semi-compact' or 'slender' section would be suitable for this design case.

For 'plastic' and 'compact' sections the member suitability is checked on the basis of general yielding. Calculate the plastic modulus S for each axis with a reduction due to axial load. These plastic moduli are used to obtain the moment capacity for each axis. The applied moments divided by moment capacity for both the x and y axes are then summed and the resulting value must not exceed unity.

$$\text{Moment capacity } M_c = p_y S \not> 1.2 \ p_y Z \qquad \text{(code 4.2.5)}$$

where p_y is the steel design strength
  S is the plastic modulus, reduced due to axial load (and shear stress, if any)
  Z is the elastic modulus.

The member suitability will be shown by the formula:

$$\left(\frac{M_x}{M_{rx}}\right)^{z_1} + \left(\frac{M_y}{M_{ry}}\right)^{z_2} \not> 1 \quad \text{(plastic \& compact sections only)} \quad \text{(code 4.8.2)}$$

$z_1 = $ a constant, taken as 2.0 for I and H sections, 5/3 for solid and hollow sections, 1.0 for all other sections, and $z_2 = 1.0$ except for solid and closed hollow sections, where 5/3 should be used.

M_x and M_y are the applied moments in the x and y axes respectively, and M_{rx} and M_{ry} are the corresponding reduced moment capacities in the presence of axial load. Reduced moment capacities are obtained from published tables (BCSA/Constrado handbook) and Section 2.3.4. For a simplified approach for compact and semi-compact sections the member is checked on the basis of yield at the extreme fibres. Thus

$$\text{Moment capacity } M_c = Z p_y$$

and member suitability is given by the formula:

$$\frac{F}{A_e p_y} + \frac{M_x}{M_{cx}} + \frac{M_y}{M_{cy}} \not> 1$$ (code 4.8.2)

where F is the axial load,

  M_{cx} is the moment capacity about the xx axis in the absence of axial load
  M_{cy} is the moment capacity about the yy axis in the absence of axial load.

2.4.3 Tension member design examples

Example 2.8

A T section cut from a UB (serial size T102 × 152 × 13kg/m, grade 43), is loaded in tension with 75 kN. The connections at each end of the T are identical, being two M20 bolts (grade 4.6) through the web.

  Check the suitability of the section for the applied tension.

SOLUTION
Applied tension on member = 75.0 kN
Gross area of section   = 15.7 cm²
Tension member attached through the flange, consider a reduced area.
Effective area $A_e = A(3a_1/(3a_1 + a_2))$ (code 4.6.3.1)
  a_1 = sectional area of connected leg   (code 3.3.2)
  a_2 = sectional area of unconnected leg.
  A = total area of the tee shape − holes
  $a_1 = 6.8 \times 102 - 2 \times 22 \times 6.8 \text{(holes)} = 394.4 \text{ mm}^2$
  $a_2 = 145.2 \times 5.8$ $= 842.1 \text{ mm}^2$
TOTAL NET AREA $= 1236.5 \text{ mm}^2$
  $A_e = 1236.5 (3 \times 394.4)/(3 \times 394.4 + 842.1) = 722.1 \text{ mm}^2$
  $p_y = 275 \text{ N/mm}^2$ (code Table 6)
  $P_t = 722.1 \times 275/10^3 = 198.5 \text{ kN} > 75 \text{ kN}$
The 102 × 152 × 13 T is adequate

Example 2.9

A structural design includes a beam tension member subject to moments about both axes. Given the design criteria below, check the suitability of the member.

Beam: 203 × 203UC46 grade 43 steel
Moment connections at each end of member, fully welded
$M_{xx} = 40.0$ kNm (ultimate)
$M_{yy} = 21.0$ kNm (ultimate)
Axial tension = 400 kN (factored)

SOLUTION

p_y for grade 43 steel = 275 N/mm² (steel thickness < 16 mm)

Beam section = 203 × 203UC46 – check for compact section   (code Table 6)

$\varepsilon = (275/275)^{1/2} = 1.0$ (code Table 7)

$b = 203.2/2 = 101.6$ mm $T = 11.0$ mm

$b/T = 101.6/11.0 = 9.24 < 9.5$ for flange outstand of compact section   (code Table 7)

$d/t = 160.8/7.3 = 22.02 < 98$ for web of compact section

Section is thus compact

Moment capacity

$M_c = S_x p_y$ as section is compact

   $S_x = 497.4$ cm³

   $S_y = 230.0$ cm³

$M_{cx} = 497.4 \times 10^3 \times 275/10^6 = 136.8$ kNm

$M_{cy} = 230.0 \times 10^3 \times 275/10^6 = 63.3$ kNm

Effective area of section

Assume that fastener holes will not reduce the area by more than 9% for which no reduction is required. When the final connection is designed, check that this requirement is still satisfied.

Area of section $A = 58.8$ cm² $= A_e$ (effective area) because of no deduction for holes.

Tension capacity $= P_t = A_e p_y = 58.8 \times 10^2 \times 275/10^3 = 1617$ kN

Unity check

$$\frac{F}{p_t} + \frac{M_x}{M_{cx}} + \frac{M_y}{M_{cy}} \not> 1 \qquad \text{(code 4.8.2)}$$

$$\frac{400}{1617} + \frac{40.0}{136.8} + \frac{21.0}{63.3} = 0.87 \not> 1$$

203 × 203UC46 is adequate (grade 43)

2.5 PLATE GIRDERS

2.5.1 Introduction

Plate girders are used in road and railway bridges, and occasionally in buildings where heavy loads and/or large spans are required. Their construction consists of steel plates which are usually welded or bolted together to form an I section.

Plate girders can be built to any size to suit design requirements, but the proportions are usually limited by web buckling. Stiffeners are used to reinforce the web, thus increasing the allowable buckling stress. For long span construction where standard beams are not suitable, the choice is generally between adding plates to standard beams or using a plate girder.

For modern road bridges, where continuous construction is used to reduce the maximum moments, a plate girder may not be required until the span exceeds about 25 m.

Generally built-up standard beams are more economical below the 25 m span.

In recent times numerous plate girders spanning 60 to 100 m have been constructed. With the introduction of automatic welding and the development of steels which permit the use of normal procedures for welding thick plates, the cost of fabrication has been considerably reduced.

2.5.2 Design strength

The design strength p_y is dependent on the grade of steel and the thickness of plate used (code Table 6). For plates of grade 43 steel the design strengths are, for plate thickness less than or equal to:

- 16 mm, $p_y = 275$ N/mm²
- 40 mm, $p_y = 265$ N/mm²
- 100 mm, $p_y = 245$ N/mm²

For welded elements in compression a reduction of 20 N/mm² of the permissible design strength is required (code 4.7.5). This reduction does not apply to elements when checked for bending, shear and connections (code 4.1.3).

For design strengths of higher grades of steel refer to Table 6 of the code.

When designing a plate girder the following points should be checked (code 4.4):

2.5.3 Bending

(a) Check limiting proportions of outstands and web   (code Table 7)
(b) At critical points check the combination of maximum
    moment and co-existent shear, and maximum shear and
    co-existent moment   (code 4.4.2.3)
(c) Check deflection limits   (code Table 5)
(d) For full lateral restraint, the positive connection of floor
    should be capable of resisting a lateral force of not less than
    1% of the maximum force in the compression flange   (code 4.2.2)
(e) Check for local buckling   (code 4.5.1.1(a))
(f) For loads that are applied thro flange to web check for
    bearing and buckling   (code 4.5)

2.5.4 Shear

The shear force should not exceed the shear capacity P_v of the web, where $P_v = 0.6 p_y A_v$ and A_v is the shear area which, for a plate girder, is td, web thickness multiplied by web depth. If the shear capacity is exceeded, the web requires thickening, or part of the shear can be carried by a triangulated system of diagonal stiffeners. If the shear force exceeds 60% of the shear capacity (i.e. average shear stress exceeds $0.36 p_y$) then the moment capacity will be reduced.

The shear buckling resistance equates to the shear which an unstiffened web carries without causing a buckling failure.

The shear buckling resistance is equal to the shear capacity based on yield, when the web depth to thickness ratio $d/t = 63\varepsilon$ or, generally, $63\sqrt{(275/p_y)}$.

If the web ratio d/t exceeds 63ε (code Table 7) it is classified as a 'thin' web. Although the buckling resistance of the web may be satisfactory, a check should be made to ensure that the web capacity based on yield is not exceeded.

2.5.5 Lateral torsional buckling resistance for members subject to bending

For a section between adjacent lateral restraints and bending about major axis,

$$\bar{M} \not> M_b \qquad\qquad \text{(code 4.3.7.1)}$$

where M_b = buckling resistance moment and

$$\bar{M} = mM \qquad\qquad \text{(code 4.3.7.2)}$$

where m is determined from code 4.3.7.6.

$$M_b = S_x p_b \qquad\qquad \text{(code 4.3.7.3)}$$

where S_x = plastic moment about xx
and p_b = bending strength   (code 4.3.7.4)

2.5.6 Web design

2.5.6.1 Minimum thickness

The thickness of webs should meet the following requirements:

(a) MINIMUM THICKNESS FOR SERVICEABILITY (code 4.4.2.2):
Without intermediate stiffeners: $t \geqslant d/250$
With transverse stiffeners only:

(1) where stiffener spacing $a > d$:$t \geqslant d/250$
(2) where stiffener spacing $a \leqslant d$:$t \geqslant d/250\sqrt{(a/d)}$

(b) MINIMUM THICKNESS TO AVOID FLANGE BUCKLING (code 4.4.2.3):

Without intermediate stiffeners: $t \geqslant (d/250)\,(p_{yf}/345)$
With intermediate transverse stiffeners:

(1) where stiffener spacing $a > 1.5\,d$: $t \geqslant (d/250)\,(p_{yf}/345)$
(2) where stiffener spacing $a \leqslant 1.5\,d$: $t \geqslant (d/250)\,(p_{yf}/455)$

where p_{yf} is the design strength of the compression flange.

2.5.6.2 Design methods

For the design of webs there are two methods stated in the code – they are:

(a) Elastic critical
(b) Tension field action

Method (a) gives the least amount of computation, but the latter method results in a more economical web by using less material.

(a) ELASTIC CRITICAL
This method may be used for the design of unstiffened girders, and for the design of internal and end panels of a stiffened girder.

The critical shear strength q_{cr} is governed by the yield strength p_y of the steel, and the ratios d/t and a/d where $a =$ spacing of transverse web stiffeners and d and t are web depth and thickness respectively. Values of q_{cr} are obtained from code Tables 21(a) to (d) for steel grades 43 and 50.

For a transversely stiffened or unstiffened web panel the shear buckling resistance V_{cr} of a web is calculated from the formula:

$$V_{cr} = q_{cr}dt$$
(code 4.4.5.3)

(b) TENSION FIELD ACTION
The concept of basic tension field action for plate girders was proposed by Basler and Thurlimann in 1963, when they formulated a tension field model.

Tension field action occurs in a plate girder web owing to stress redistribution after web buckling occurs as a result of shear. The flanges and vertical (transverse) stiffeners divide the girder into panels. In each panel a diagonal strip of web acts as a tension member, and the vertical stiffeners as compression struts. This action is analogous to the panels of a Pratt truss (Fig. 2.22). The tension fields are assumed to be 'anchored' at the vertical stiffeners. The end panel acts as an anchor for the tension field, and one method adopted to resist the horizontal component of the anchorage forces is to introduce an end post adjacent to the end stiffener (see Fig. 2.23).

Stocky webs will carry the full shear yield stress, $p_y/\sqrt{3}$, over the entire web

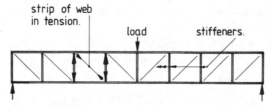

Fig. 2.22 Tension field action truss analogy.

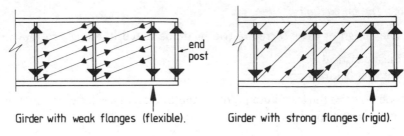

Girder with weak flanges (flexible).   Girder with strong flanges (rigid).

Fig. 2.23 Tension field action – post buckling of stiffened web panels.

depth when 'pure' shear is applied, and the applied shear force will be carried by beam action shear.

For thin webs the applied shear force is resisted by beam action shear until web buckling stress occurs, after which the tension field action resists any residue of shear force.

The total applied shear which a panel can sustain is therefore the addition of beam action shear plus the tension field contribution. When the flanges are completely flexible 'basic' tension field action occurs, and the tension field 'hangs' between the stiffeners. In this case the flange area is not utilised (Fig. 2.23). This method may be used for the design of internal and end panels providing the end panels are designed as stated in code 4.4.5.4.

The shear resistance of a stiffened panel is obtained from

$$V_b = q_b dt$$

If the flanges in the panel are not fully stressed the shear resistance may be increased to:

$$V_b = (q_b + q_f \sqrt{K_f}) dt \text{ but } \leqslant 0.6 \, p_y dt$$

where q_b = basic shear strength from code Tables 22(a) to (d)

q_f = the flange dependent shear strength factor from code Tables 23(a) to (d)

$K_f = (M_{pf}/4M_{pw})(1 - f/p_{pf})$

M_{pf} = plastic moment capacity of smaller flange about its own horizontal axis

f = the mean longitudinal stress in smaller flange due to moment and/or axial load

p_{yf} = design strength of the flange

M_{pw} = plastic moment capacity of web about its own horizontal axis.

If the shear does not exceed the critical shear strength, then tension field action need not be considered. The shear buckling resistance with and without tension field action varies according to the stiffener spacing and the d/t ratio of the web.

2.5.7 Moment capacity for slender webs (non-compact)

If the web of a girder is sufficiently thick, the moment capacity limit depends on whether the compression flange is compact, semi-compact or slender. A compression flange which is stressed to yield at its extreme fibres without local buckling is designated as semi-compact. To attain the plastic moment capacity of the section, the strain at the extreme fibres increases at a constant stress, until redistribution of bending stress occurs, from the flange into the web. A compact compression flange is required to withstand this strain without local buckling occurring.

The assumptions made are that there is no axial load or shear force, and that the web is sufficiently thick in order to prevent local buckling of the web plate. If the web is not thick enough to prevent local buckling it is called a 'slender' web. This concept is not the same as for a 'thin' web, which requires a check for shear buckling. A slender web prevents the full moment capacity being reached, because of its tendency to suffer from local buckling.

The tabulated limiting values in the code for compact, semi-compact and slender webs and compression flanges are provided to indicate which criterion will govern in a particular design case. Application of additional moment to a web which has commenced local buckling results in redistribution of stress from part of the web in compression into the compression flange.

The combined effect of direct stresses due to moment and axial load with the effects of shear should be considered. It should be sufficiently accurate to combine them as if they were all elastic critical stresses, and the combination of these effects with direct stresses due to loads applied through the flange to the web should be investigated.

When a web is slender, but the flanges are plastic, compact or semi-compact, that is $d/t \geqslant 63\varepsilon$, the code gives three methods for calculating the moment capacity (code 4.4.4.2):

(1) Assume that the flanges resist the moment and axial load and the web designed for shear only
(2) Assume that the whole section resists the moment and axial load, and the web designed for combined shear and longitudinal stresses
(3) A proportion of the loading is resisted by method (2) and the remainder of the loading by method (1) with the web designed accordingly.

2.6 INTERMEDIATE TRANSVERSE WEB STIFFENERS

These stiffeners may be on one or both sides of the web, and their spacing should comply with code 4.4.2 dependent on the thickness of the web.

2.6.1 Outstand

The outstand of the stiffeners (Fig. 2.24) from the face of the web should not exceed $19t_s\sqrt{(275/p_y)}$, unless the outer edge is continuously stiffened.

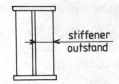

Fig. 2.24 Transverse stiffener.

If the outstand exceeds $13t_s\sqrt{(275/p_y)}$ then design the stiffeners using a core section with an outstand of $13t_s\sqrt{(275/p_y)}$ (code 4.5.1.2)

2.6.2 Minimum stiffness

Transverse web stiffeners that are not subject to external loads or moments should have a second moment of area I_s about the centreline of web as follows (code 4.4.6.4):

$$I_s \geqslant 0.75\,dt^3 \text{ for } a \geqslant d\sqrt{2}$$

and

$$I_s \geqslant \frac{1.5\,d^3t^3}{a^2} \text{ for } a < d\sqrt{2}$$

where d = web depth

t = min. required web thickness for spacing a using
tension field action   (code 4.4.5.4.1)

a = actual stiffener spacing

2.6.3 Buckling

The stiffeners should be checked as struts, and are required to resist a force equal to the difference between the shear in the web adjacent to the stiffener and the shear resistance of the unstiffened web.

Stiffeners that are not subject to external loads or moments should be checked for a stiffener force.

$$F_q = V - V_s \leqslant P_q \qquad\qquad \text{(code 4.4.6.6)}$$

where V = max. shear adjacent to stiffener

V_s = shear buckling resistance of web panel designed without
using tension field action.   (code 4.4.5.3)

For the requirements of minimum stiffness and buckling for intermediate stiffeners subjected to external loads the reader is referred to the code clauses 4.4.6.5 and 4.4.6.6.

2.6.4 Connection to web   (code 4.4.6.7)

Stiffeners that are not subject to external loading should be connected to the web to withstand a shear force between each stiffener and the web (in kN/mm^2)

of not less than

$$t^2/8b_s$$

where b_s = stiffener outstand (mm)
t = web thickness (mm)

If the stiffeners are subjected to external loading the shear between the web and the stiffener should be added to the value obtained from $t^2/8b_s$.

The stiffeners should extend to the compression flange but are not necessarily connected to it.

2.7 STIFFENER DESIGN FOR BEARING AND BUCKLING

This deals with the design of webs of beams and girders subject to loading acting through the flange parallel to the plane of the web. When the web without stiffeners proves to be inadequate, then stiffeners are provided to cover the following requirements (code 4.5.1.1).

(a) To prevent local buckling of web due to concentrated load (load carrying stiffeners)
(b) To prevent local crushing of web due to concentrated load (bearing stiffeners)
(c) To prevent buckling of a slender web due to shear (intermediate stiffeners)
(d) To provide torsional restraint at girder supports (torsion stiffeners)
(e) To provide local reinforcement of web in shear and bearing (diagonal stiffeners)
(f) To transmit tensile loads applied to a web through a flange (tension stiffeners)

A stiffener may perform one, or a combination of, the above-mentioned functions and its design shall comply with all the requirements of those functions.

2.7.1 Stiff bearing length   (code 4.5.1.3)

To determine the stiff bearing length b the dispersion of load through a steel bearing is generally taken as 45 degrees as illustrated in Fig. 2.25.

2.7.2 Buckling resistance   (code 4.5.1.5)

The buckling resistance of a stiffener is based on the compressive strength of a strut, whose radius of gyration is taken about the axis parallel to the web. The effective section is taken as the full or core area of the stiffener plus an effective length of web on each side of the centreline of the stiffener limited to 20 times web thickness.

For calculating buckling resistance, the effective length (L_e) for intermediate transverse stiffeners should be taken as $0.7L$, where L = stiffener length.

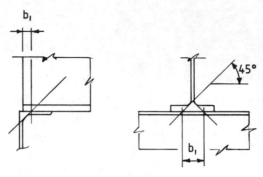

Fig. 2.25

If a stiffener is attached to a welded section, a reduction of 20 N/mm² should be used for the design strength (code 4.7.5).

The effective length for load carrying stiffeners is dealt with in Section 2.8.1.

2.8 LOAD CARRYING STIFFENERS   (code 4.5.2)

To prevent local buckling see Section 2.2.5.

2.8.1 Buckling resistance

Where a compressive force applied through a flange exceeds the buckling resistance of an unstiffened web (Fig. 2.26), load carrying stiffeners should be provided. The buckling resistance is:

$$P_w = (b + n_1)tp_c$$

where b = stiff length of bearing   (code 4.5.1.3)
   n_1 = length derived by projection at 45° through half depth of girder
   t = web thickness
   p_c = compressive strength   (code 4.5.1.4 and Table 27(c))

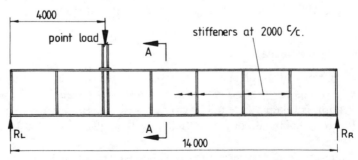

Fig. 2.26 Plate girder design, example.

The buckling resistance of a load carrying stiffener based on the compression resistance of a strut shall be not be less than the applied load F_x on the stiffener (code 4.5.1.5). The effective length $L_e = 0.7 \times$ length if the rotation between web and flange is restrained (if unrestrained use $L_e = L$).

2.8.2 Bearing check

Load carrying web stiffeners shall have an area of

$$A > 0.8 F_x / p_{ys} \qquad \text{(code 4.5.4.2)}$$

where F_x = applied load or reaction
  A = area of stiffener in contact with flange
  p_{ys} = design strength of stiffener.

2.9 BEARING STIFFENERS (see Section 2.2.6)

Bearing stiffeners shall be provided when a load applied through a flange exceeds the local capacity of the web at its connection to the flange. The limit on the unstiffened web is:

$$(b + n_2) t p_y \qquad \text{(code 4.5.3)}$$

where b = length of stiff bearing
  n_2 = length dispersed through flange at a slope of 1:2.5
  t = web thickness
  p_y = web design strength.

2.10 PLATE GIRDER — DESIGN EXAMPLE

Example 2.10

(Refer to Figs 2.26, 2.27 and 2.28)

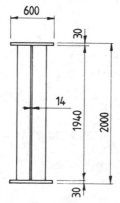

Fig. 2.27 Section A–A.

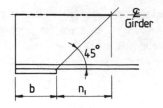

Fig. 2.28 Plan of girder.

Use grade 43 steel.
Loading

	Load factor	kN
Uniformly distributed	Imposed $550 \times 1.6 =$	880
	Dead $\quad 600 \times 1.4 =$	840
		1720 kN

Point load	Imposed $400 \times 1.6 =$	640
	Dead $\quad 500 \times 1.4 =$	700
		1340 kN

Reactions

$$R_L = 1340 \times \frac{10}{14} + \frac{1720}{2} = 1817\,\text{kN}$$

$$R_R = 1340 \times \frac{4}{14} + \frac{1720}{2} = 1243\,\text{kN}$$

$$M_{max} = 1817 \times 4 - \frac{1720}{14} \times \frac{4^2}{2} = 6285\,\text{kN m}$$

Try 2000 deep girder: Flanges 600×30 mm
Web 1940×14 mm

Assume that the moment is resisted by the flanges and subject to uniform stress p_y and the web designed for shear only. The compression flange is restrained against lateral torsional buckling.

Flange check:

Design strength $p_y = 265$ N/mm^2 (for plate thickness $> 16 < 40$ mm) (code Table 6)

$b/T = 293/30 = 9.7 < 13 \sqrt{(275/265)} = 13.2$ (code Table 7)

The flanges are therefore semi-compact.

$$\text{Stress in flanges} = \frac{6285 \times 10^6}{1970 \times 600 \times 30} = 177\,\text{N/mm}^2 < 265$$

Shear:

Shear capacity $P_v = 0.6 p_y A_v = 0.6 \times 265 \times 194 \times 14 = 4318$ kN

Max. shear force $= 1817$ kN

If $F_v \leqslant 0.6 p_v$, then, for a semi-compact section,

moment capacity $M_c = p_y Z = 265 \times 43449 \times 10^{-3} = 11514$ kNm

Max. $M = 6285$ kNm

Web $d/t = 1940/14 = 139$.

With stiffeners at 2000 mm centres $a/d = 2000/2000 = 1$
Then critical shear strength $q_{cr} = 89 \text{ N/mm}^2$ (code Table 21(a))

Average shear stress $= \dfrac{1817 \times 10^3}{1940 \times 14} = 89 \text{ N/mm}^2$

Minimum web thickness (for serviceability)   (code 4.4.2.2)
For transverse stiffeners where spacing $a \leqslant d$

$$t \geqslant \frac{d}{250}\sqrt{\frac{a}{d}} = \frac{1940}{250}\sqrt{1} = 7.8 \text{ mm}$$

actual web $= 14$ mm thick.
Minimum web thickness to avoid flange buckling: (code 4.4.2.3)
with intermediate transverse stiffeners where spacing $a \leqslant 1.5d$

$$t \geqslant \frac{d}{250}\sqrt{\frac{p_{yf}}{455}} \leqslant \frac{1940}{250}\sqrt{\frac{245}{455}} = 5.7 \text{ mm}$$

where $p_{yf} =$ design strength of compression flange.
Shear buckling resistance of web panel, V_{cr}
$V_{cr} = q_{cr}dt = 89 \times 1940 \times 10^{-3} \times 14 = 2417 \text{ kN}$ (code 4.4.5.3)
Web buckling resistance:
Buckling check: assume 80 mm stiff bearing
Slenderness, $\lambda = 2.5d/t = 2.5 \times 1940/14 = 346$ (code 4.5.2.1)
Buckling resistance $P_w = (b + n_1)tP_c$ (code Table 27(c))
$(80 + 1000)\,14 \times 15.5 \times 10^{-3} = 234 \text{ kN} < 1817 \text{ kN}$
Provide load carrying stiffeners.   (code 4.5.2)

Bearing check:
Local capacity of web at flange connection $= (b + n_2)tp_{yw}$ (code 4.5.3)
Bearing capacity $(80 + 30 \times 2.5)14 \times 0.275 = 596 \text{ kN} < 1817 \text{ kN}$
$\therefore$ bearing stiffener capacity required $= R_L - 596 = 1221 \text{ kN}$ (code 4.5.5)
Since the area of stiffener in contact with the flange, $A > 0.8F_x/P_{ys}$ (code 4.5.4.2)
design stiffeners to carry applied load.   (code 4.5.4.2)

Required area $= \dfrac{1817 \times 0.8}{0.255} = 5700 \text{ mm}^2$

Maximum outstand $= 13t_s\,\varepsilon = 13t_s$ as $\varepsilon = (275/275)^{\frac{1}{2}}$ (code 4.5.1.2)
Area $= dt_s$. Assuming $t_s = d/24$, area $= d^2/24$
then $d = \sqrt{(5700 \times 24)} = 370 \text{ mm}$
and $t_s = 370/24 = 15.4 \text{ mm}$

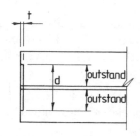

Fig. 2.29 Web buckling check.

Make stiffener $400\,\text{mm} \times 16\,\text{mm} = 6400\,\text{mm}^2$ (2No.)

Second moment of area $I_s = (40 \times 1.4)^3 \times \dfrac{1.6}{12} - 1.4^3 \times \dfrac{1.6}{12} = 23415\,\text{cm}^4$

for $a \leqslant 2d$, $I_s \geqslant \dfrac{1.5d^3t^3}{a^2}$

$$= \frac{1.5 \times 194^3 \times 1.4^3}{200^2} = 751\,\text{cm}^4 \qquad \text{(code 4.4.6.4)}$$

Bearing capacity of stiffened web
$A = $ stiffened area $+ 20 \times$ web $\times t^2$.

With flanges restrained $\lambda = 0.7 \times \dfrac{194}{15(r_y)} = 9.0$

where $r_y = \sqrt{\dfrac{23415}{1.4 \times 40 + 1.4 \times 20 \times 1.4}} = 15\,\text{cm}$

Capacity $= (16 \times 400) + (280 \times 14) \times 0.2 = 2064\,\text{kN}$
$\qquad\qquad\qquad\qquad\qquad\qquad\qquad > 1817\,\text{kN}$

Load on weld $1817 - 477 = 1340\,\text{kN}$.

Load per mm of weld $= \dfrac{1340 \times 10^3}{1940 \times 4} = 173\,\text{N/mm}$

For intermediate transverse stiffener, required minimum shear
$= t^2/8b = 14^2/8 \times 200 = 123\,\text{N/mm}$ $\qquad\qquad$ (code 4.4.6.7)
where $b = $ stiffener outstand
Fillet weld design strength $p_w = 215\,\text{N/mm}^2$ $\qquad\qquad$ (code Table 36)
Use 6 mm weld – then $0.7 \times 6 \times 215 = 900\,\text{N/mm}$

2.11 PROBLEMS

(All loads are unfactored unless otherwise stated)

2.1 A simply supported beam has a clear span of 7 m and a central point imposed load of 20 kN. Calculate the minimum moment of inertia (I) to limit the deflection to the permissible value.

2.2 A gable column of 5 m height with pinned ends is subjected to uniform wind loading of 6 kN/m. Calculate the minimum I value to satisfy deflection limits.

2.3 A $120 \times 120 \times 10$ angle brace (grade 43 steel) is subjected to axial tension of 56 kN dead load. Connections at each end are identical, being 2 No.M20 bolts (grade 4.6) in line with the load. Check the suitability of the angle section and the bolts for the applied load.

2.4 A pin ended column 305×165 UB 40, length 5 m (grade 50 steel) is subjected to axial tension and bi-axial bending as noted below. Check the suitability of the member.

$$\text{Imposed loads} \begin{cases} \text{Axial tension} = 100\,\text{kN} \\ M_y \qquad\qquad = 5\,\text{kNm} \\ M_x \qquad\qquad = 28\,\text{kNm} \end{cases}$$

2.5 Design a column using the following design criteria:
Unrestrained length = 7 m.
One end fixed, other end pinned. These end conditions apply to both xx and yy axes. Compressive loads – 200 kN axial dead load; 1000 kN live load. Use grade 50 steel.

2.6 A simply supported beam 457 × 152 UB 52 (grade 50 steel) is subjected to an end reaction of 178 kN dead load.
Check (a) the average shear stress at beam end
      (b) the elastic shear stress of the beam at the neutral axis.
State whether the beam is adequate to resist the shear at this location.

2.7 Design a fixed ended beam for the following criteria:
Span = 8 m
Max. unrestrained length of top flange = 4 m
Loading = 5 kN/m dead load (including self weight)
              10 kN/m live load
'Framed' end connection.

REFERENCES

1. Various authors (1978) *The Background to the New British Standard for Structural Steelwork*, Imperial College, London.
2. McGuire, W. (1968) *Steel Structures*. Prentice-Hall, Englewood Cliffs, N.J.
3. Lambert Tall (1974) *Structural Steel Design*. Ronald Press Company, N.Y.
4. Lambert, F. W. (1973) *Structural Steelwork*. MacDonald & Evans, London.
5. Blodgett, G. W. (1966) *Design of Welded Structures*. James F. Lincoln Arc Welding Foundation.
6. Robertson, A. (1925) *The Strength of Struts*. Inst. C.E. Selected Engineering Paper No. 28, Inst. Civ. Engrs.
7. Godfrey, G. B. (1962) The allowable stresses in axially loaded steel struts. *Structural Eng.* **40**, No. 3.
8. Young, B. W. (1975) Buckling of axially loaded welded steel columns. *Structural Eng. May.*
9. *Steel Plated Structures – An International Symposium* (1977) (Eds P. J. Dowling, J. E. Harding and P. A. Frieze) Crosby Lockwood Staples, London.
10. Chan, C. and Ostapenko, A. (1969) *Ultimate Strength of Plate Girders Under Shear*. Lehigh Univ. Fritz Eng. Lab. Report 328.7, August.
11. Nethercot, D. A. (1972) Factors affecting the buckling stability of partially plastic beams. *Proc. Inst. Civ. Engrs.* **53**, September.

12. Nethercot, D. A. (1974) Buckling of welded beams and girders. *Pubs. Int. Assoc. for Bridge and Struct. Eng.* **34**, March.
13. Nethercot, D. A. (1976) Lateral buckling approximations for elastic beams. *Structural Eng.* No. 6, June.
14. Needham, F. H. (1977) The economics of steelwork design. *Structural Eng.* September.
15. Smith and Riddington (1978) The design of masonry infilled steel frames for bracing structures. *Structural Eng.* **56B**, March.

Steelwork connections **3**

3.1 DEFINITION AND PHILOSOPHY

A connection may be defined as the component parts used to join together elements or members in a structure.

The function of a connection is to transmit co-existent forces and moments between members at the joints, or 'node' points. Connections are usually categorized as pinned or fixed, depending on their ability to transmit moments.

A pinned connection is able to transmit shear and axial load, and a fixed connection will transmit shear, axial load and moments. In practice, a pure pinned or fully fixed condition is not necessarily achieved, but the terms are used to describe the connection's prime function in the structural design.

Where plastic theory is used for design, a fixed connection must develop the plastic moment at the point considered, which may be less than the full plastic moment.

Connections may consist of assemblies of plates and angle sections, bolted, riveted, or welded together, including direct welding between members, or a combination of bolting and welding. Riveting has largely been replaced by the use of welding or high strength friction grip bolting. (In the past, rivets were used for their higher load carrying capacity, and non-slip properties, where ordinary bolts would have slipped by at least 1 mm, that is, half the hole tolerance.) All welded connections must be restricted to transportable structural assemblies, unless site welding is envisaged. It is now common practice to weld component parts of connections in the fabrication shop, and adopt bolting assembly on site.

Combination of welding and other fasteners in a given shear plane for a connection should be avoided, but if there is no alternative, then HSFG bolts or rivets should be used. Ordinary bolts in clearance holes should not be used where reversal of stress, or vibration, occurs.

End connections of a member influence the effective length assumed for the member design. Effective lengths have already been defined in Chapter 2.

3.2 BOLT TYPES – GENERAL (see Table 3.1)

Bolts are used to join two or more separate plates, resisting shear at the plate interface and tension in the bolt axis.

Tension on the connected plate, eccentric to the bolt axis, will induce a prying action in the bolt, where prying is caused by an additional force on the bolt,

Table 3.1. Bolt data

Iso-Metric bolts grade 4.6 and 8.8 to BS4190 (washers to BS4320)

Bolt dia. mm	Washer O/d (Form E) mm	O/a bolt corners mm	Allowance to add to grip mm	Nominal hole dia. mm	Tensile stress area mm²
M16	30	27.7	19	18	157
M20	37	34.6	23	22	245
(M22)	39	36.9	25	24	303
M24	44	41.6	28	26	353
(M30)	56	53.1	33	33	561
M33	60	57.7	36	36	694
M36	66	63.5	40	39	817

Iso-Metric bolts

Bolt dia. mm	Grade 4.6 (Black, mild steel) Proof load kN		Ultimate load kN	Grade 8.8 (high tensile) Proof load kN		Ultimate load kN
M16	34.8		61.6	89.6		123
M20	54.3		96.1	140.0		192
(M22)	67.3	not normally tightened to Proof L.	118.8	173.0	not normally tightened to Proof L.	238
M24	78.2		138.0	201.0		277
(M30)	124.0		220.0	321.0		439
M33	154.0		271.0	396.0		544
M36	181.0		321.0	466.0		641

HSFG bolts to BS4395

Bolt dia. mm	General Grade Part 1 t.s.a. mm²	Proof L. kN	Yield L. kN	Ult L. kN	Typical Dimns all HSFG Bolts. Washer* O/d mm	O/a blt crs mm	a mm	nom. hole mm
M16	157	92.1	99.7	130	37	31.2	26	18
M20	245	144.0	155.0	203	44	36.9	30	22
M22	303	177.0	192.0	250	50	41.6	34	24
M24	353	207.0	225.0	292	56	47.3	36	27
M27	459	234.0	259.0	333	60	53.1	39	30
M30	561	286.0	313.0	406	66	57.7	42	33
M33	–	–	–	–	75	63.5	45	36
M36	817	418.0	445.0	591	85	69.3	48	39

HSFG bolts to BS4395

Bolt dia. mm	Higher Grade Part 2. t.s.a. mm²	Proof L. kN	Yield L. kN	Ult. L. kN	Higher Grade Part 3 waisted shank t.s.a. mm	Proof L. kN	Yield L. kN	Ult. L. kN
M16	157	122.2	138.7	154.1	123	95.4	108.5	120.6
M20	245	190.4	216.0	240.0	194	150.5	171.1	190.3
M22	303	235.5	266.9	296.5	243	188.6	214.3	238.4
M24	353	274.6	311.4	346.0	279	216.5	246.1	273.7
M27	459	356.0	405.0	450.0	369	286.3	325.5	362.0
M30	561	435.0	495.0	550.0	448	347.6	395.1	439.5
M33	694	540.0	612.0	680.0	562	436.1	495.7	551.3
M36	–	–	–	–	–	–	–	–

By courtesy of the British Standards Institution. *Flat, round. t.s.a. = tensile stress area.
a = allowance to add to grip.

illustrated in Fig. 3.11, and further described in Section 2.6. Prying should be included in the design, but within the guidelines of BS5950, prying forces may be assumed to be well inside the tension capacity of the bolt (see Section 3.6 and code 6.3.6.2.).

The dimensions and mechanical properties of bolts commonly used in the UK at present are specified by the British Standards Institution. There are three grades of structural bolts:

(1) Grade 4.6 bolts, often called 'ordinary' bolts, with grade 4 nuts, to BS4190
(2) Grade 8.8 bolts, with grade 8 nuts, usually to BS4190 tolerances ('precision' bolts are available in grade 8.8 to BS3692) (grades 10.9 and 12.9 are also available)
(3) High strength friction grip bolts to grade 8.8 or 10.9.

The bolt grade number is given by the ISO method as follows: The first number is a tenth of the ultimate tensile strength expressed in kgf/mm^2; the second number is the ratio of yield strength to the ultimate tensile strength $\times 10$; the product of the two numbers indicates the yield strength of the bolt material.

The thickness of plates secured by the bolt is termed the 'bolt grip'. Bolts (1), (2) and (3) are normally used in 'clearance' holes which are nominally 2 mm larger than the bolt shank diameter. Types (1) and (2) are used in 'bearing' type connections where shear at the plate interface is carried by the bolt bearing on to the plate. The tightening procedure consists of applying a recommended torque to the bolt in accordance with the manufacturer's instructions. (A bolt tension of about 85% proof load is usually required, using an approximate torque of $0.2 \times 85\%$ proof load $\times$ bolt dia.) An initial slip on 'bearing' type connections of 1 mm may occur.

Bolts type (3) are used for 'friction' type connections. HSFG bolts are tightened to induce a known shank tension (proof load) in order to clamp the two plates together. Shear at the plate interface is resisted by friction between the plates, caused by the bolt clamping action. A contact surface free from paint, dirt, rust, or oil may have a preliminary design slip factor of 0.45. The slip factor used should be verified by tests using the exact materials to be employed on the contact area. Faying surfaces are usually shot blasted and masked until immediately before assembly for the joint on site (slip factor = 0.5) and for export work the faying surface may be shot blasted and treated with zinc silicate (slip factor 0.35). BS5400: Part 3 gives a table of slip factors for various surfaces. The advantage of using HSFG bolts in a connection is that the plates are clamped together under the bolt prestress, and will not part until the tensile force on the plate equals the proof load in the bolts. The HSFG bolts are suitable for major connections where moment reversals occur, such as the eaves connection of a portal frame building.

The choice of bolt type depends on the selected connection and the

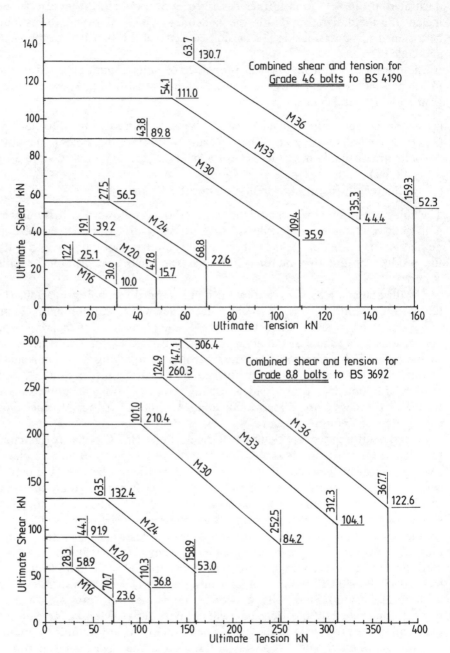

Fig. 3.1 Grade 4.6 and 8.8 bolts in clearance holes.

forces/moments to be resisted. For standard type connections bearing type bolts, grade 4.6 or 8.8, are generally specified.

With the increasing use of grade 50 steel (with a consequent reduction of lever arm for a moment connection) the selection of grade 8.8 bolts is preferable in bolt bearing situations.

High strength friction grip bolts are generally used when moment reversal or vibration occurs.

For grade 4.6 (and 8.8) combined shear and tension values, refer to graphs, Fig. 3.1.

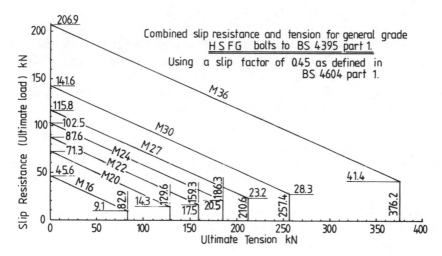

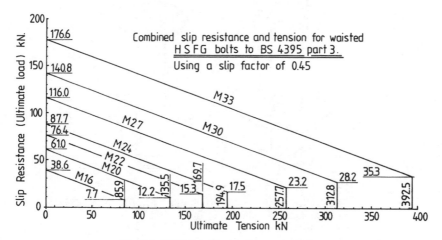

Fig. 3.2 HSFG bolts in clearance holes.

For general grade HSFG (and waisted HSFG bolts) combined slip resistance and tension values, refer to graphs, Fig. 3.2.

3.3 THE USE OF HSFG BOLTS

High strength friction grip bolts are available as listed below (see also Table 3.1).

HSFG Bolt type (BS4395)		Tightening Method (BS4604)
Part 1.	General grade HSFG bolts (Grade 8.8) Fig. 3.3(a). Nuts grade 10.	Torque control; part turn of nut; any other proven method.
Part 2.	Higher grade HSFG bolts (Grade 10.9) parallel shank. Nuts grade 12.	Part turn method not allowed; use torque control or any other proven method.
Part 3.	Higher grade HSFG bolts (Grade 10.9) waisted shank Fig. 3.3(b). Nuts grade 12.	Torque control method not allowed; use any other proven method.

The reliable use of HSFG Bolts depends on the method of tightening the individual bolt. Each bolt must be stressed to proof load in the unloaded connection. Three methods of tightening the bolt have been evolved; these are known as part turn of the nut, torque control, and gap closure. The first two methods are less accurate because they can be influenced by friction on the bolt threads. With the gap closure method, proof load in the bolt is achieved by the

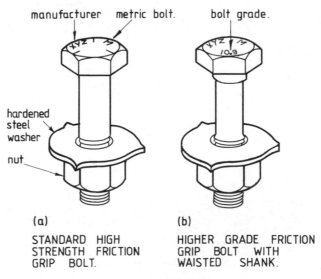

(a)
STANDARD HIGH
STRENGTH FRICTION
GRIP BOLT.

(b)
HIGHER GRADE FRICTION
GRIP BOLT WITH
WAISTED SHANK.

Fig. 3.3 HSFG bolts.

COOPER & TURNER
CORONET LOAD INDICATOR
    Fit with projections touching
underside of bolt head, or for
under nut fitting, place load
indicator with projections facing
away from steelwork, then fit on
washer & nut. After tightening
check gap specified by C & T.

(illustrated by permission of
Cooper and Turner Ltd.)

Fig. 3.4 High-strength friction grip bolting – gap closure method.

use of a 'load indicator' device, so that when the bolt is tightened a specified gap is achieved. There is a version of this concept on the UK market – namely the 'coronet load indicator' made by Cooper & Turner Ltd (Fig. 3.4). This device relies on crushing and partial shearing of metal projections in order to produce a measurable gap for inspection. When load indicators are used inspection of HSFG bolted joints should include periodic checks with a bolt meter (that is, a special hydraulic load cell with a central hole).

Assembly and tightening of HSFG bolts must follow a set procedure in order to guarantee their effectiveness. *After* the bolts have been inserted into their holes, the projecting threads may be lightly oiled to obviate excessive thread friction, taking care not to spill oil on the plates. Where there are more than four bolts in a joint, the bolts should be tightened in a staggered pattern working from the centre of the joint, outwards. If any bolt is slackened off after tightening, it *must* be discarded. When using the gap closure HSFG bolt system, the experienced operator will not need to check every bolt gap, being able to judge the correct setting in general by the feel or sound of the wrench. Bolts tightened first may need to the retightened because of bedding down.

HSFG general grade bolts to BS4395: Part 1 may be used in locations subject to axial tension, whereas higher grade HSFG bolts to BS4395: Part 2 are only suitable in shear locations. HSFG higher grade bolts (with waisted shanks), to BS4395: Part 3 are suitable for axial tension and, because of their greater ductility, are often used in fatigue connections.

3.4 WELDS

Welding is the process of joining steel elements by raising the local temperature in order effectively to fuse the two pieces of steel together. This basic definition

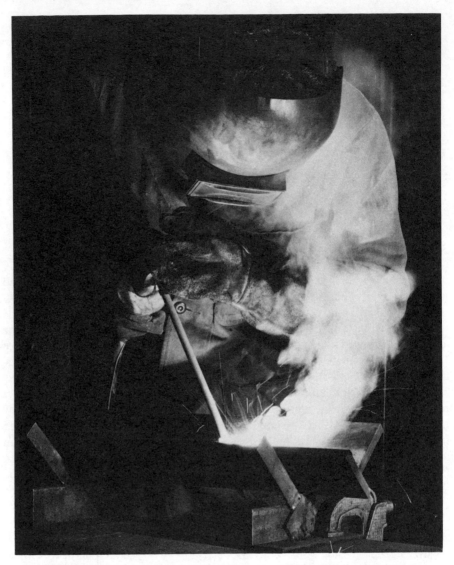

Fig. 3.5 Manual metal arc welding. (Courtesy of The Welding Institute.)

also applies to a variety of processes such as laser welding, pressure welding and fusion welding. The majority of steelwork welds are formed by the fusion welding process whereby steel parts are joined by molten metal deposited from an electrode arcing between the rod and the work to be welded.

A selection of weld types and symbols is illustrated in Fig. 3.6. The most typical types of weld used for structural purposes are the 'fillet' weld and the

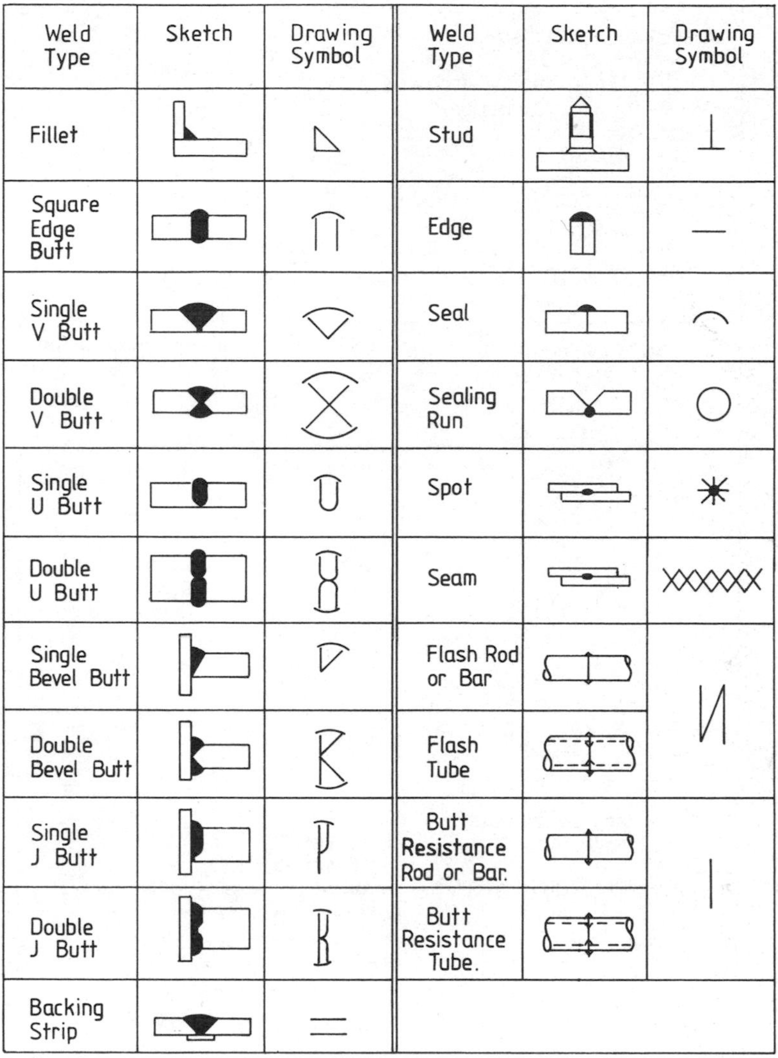

Weld Type	Sketch	Drawing Symbol	Weld Type	Sketch	Drawing Symbol
Fillet		△	Stud		⊥
Square Edge Butt		⊓	Edge		—
Single V Butt		▽	Seal		⌒
Double V Butt		⧓	Sealing Run		○
Single U Butt		⋃	Spot		✳
Double U Butt		8	Seam		XXXXXX
Single Bevel Butt		▷	Flash Rod or Bar		
Double Bevel Butt		K	Flash Tube		Ν
Single J Butt		⌐	Butt Resistance Rod or Bar.		
Double J Butt		匚	Butt Resistance Tube.		|
Backing Strip		=			

Fig. 3.6 Welding symbols.

'butt' weld. Fillet welds are roughly triangular in shape, deposited in a 90° – 120° corner. Butt welds are made by filling a gap between two adjacent plates. The strength of the weld steel is generally higher than that of the parts joined. Butt welds are deemed to have the strength of the cross-sectional area of the parts they join. Run-on and run-off pieces are used at flange butt welds to ensure the correct weld profile for the intended flange width. Fillet welds are designed to

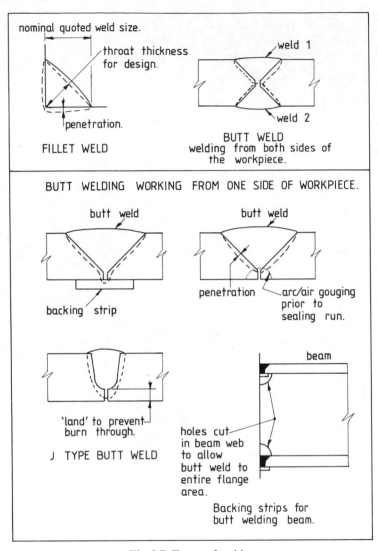

Fig. 3.7 Types of weld.

carry loads through shear taken on the throat area as shown in Fig. 3.7 with strength values as given in Fig. 3.8.

The choice of weld type depends on the required strength of connection, weld location and comparative cost.

Fillet welds are used in most situations and are generally applied down-hand or sideways. Overhead welding should be avoided unless absolutely necessary. Typical welding clearances are illustrated in Fig. 3.9.

Grade of Steel in B.S. 4360	Weld Size in mm.	Electrode Strength to B.S. 639.		
		E 43	E 51	E 51 *
40 or 43	5	0.75	0.75	
	6	0.90	0.90	
	8	1.20	1.20	
	10	1.51	1.51	
WR 50 or 50	5	0.75	0.89	
	6	0.90	1.07	
	8	1.20	1.43	
	10	1.51	1.79	
55	5		0.89	0.96
	6		1.07	1.16
	8		1.43	1.54
	10		1.79	1.93

Fig. 3.8 Strength of fillet welds in kN mm run for plates at 90°.

When butt welding is required the type of plate preparation should be stated. The welding symbols shown in Fig. 3.6 indicate some of the various plate preparations that are required for butt welds.

The root opening is the separation between members to be joined. The opening is used for electrode accessibility to the base or root of the joint. The smaller the angle of the bevelled plate the larger the root opening required to obtain a satisfactory fusion at the root.

When joining plates with different thicknesses the thicker plate is usually tapered to a slope of 1 in 4.

3.5 COLUMN BASEPLATES

The design assumptions for a given structure will dictate the type of column base required. Where a column base is designed as pinned (will not resist moment), it generally consists of a simple slab baseplate welded to the column (Fig. 3.10a). If the base is designed as 'fixed' (i.e. rigid), it is likely that some variation of Fig. 3.10b to Fig. 3.10d would be employed. Type 3.10c is more often used for bolts over 30 mm diameter, introducing a certain amount of

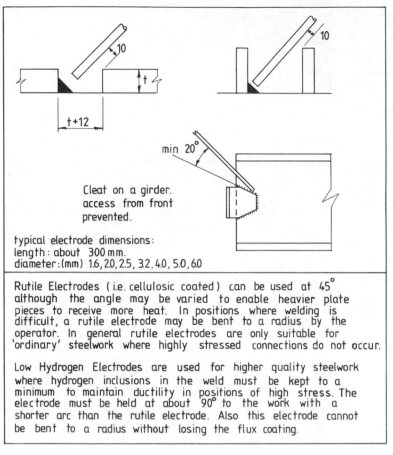

Cleat on a girder.
access from front
prevented.

typical electrode dimensions:
length: about 300 mm.
diameter:(mm) 1.6, 2.0, 2.5, 3.2, 4.0, 5.0, 6.0

Rutile Electrodes (i.e. cellulosic coated) can be used at 45°
although the angle may be varied to enable heavier plate
pieces to receive more heat. In positions where welding is
difficult, a rutile electrode may be bent to a radius by the
operator. In general rutile electrodes are only suitable for
'ordinary' steelwork where highly stressed connections do not occur.

Low Hydrogen Electrodes are used for higher quality steelwork
where hydrogen inclusions in the weld must be kept to a
minimum to maintain ductility in positions of high stress. The
electrode must be held at about 90° to the work with a
shorter arc than the rutile electrode. Also this electrode cannot
be bent to a radius without losing the flux coating.

Fig. 3.9 Welding clearances.

ductility into the connection due to the unanchored length of bolt. The nature
of connection 3.10(c) also demands accurate HD bolt positioning on site.

The code suggests that column bases be designed by an empirical method
where the nominal pressure is determined on the assumption of linear pressure
distribution. In fact if the base was allowed to deform in a plastic fashion, the
loads would not be properly transmitted to the base. Therefore, the base should
be designed elastically, that is, using elastic modulus Z values. For biaxial
moments, the bolts and baseplate should be designed in the same way as
reinforced concrete at that particular plane. The weld at the contact surface
between column and baseplate should be checked for compression as well as
tension.

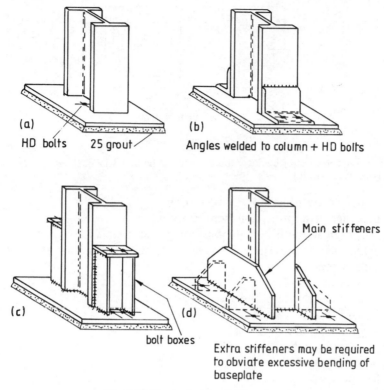

(a)

HD bolts   25 grout

(b)

Angles welded to column + HD bolts

Main stiffeners

(c)

bolt boxes

(d)

Extra stiffeners may be required
to obviate excessive bending of
baseplate

Fig. 3.10 Typical column bases.

3.5.1 Procedure for empirical design of a slab baseplate for axial load only – pinned connection

1. Determine axial load and shear at column base (ultimate forces).
2. Decide on number and type of holding down bolts to resist shear and tension. Regular number of bolts to be arranged symmetrically near corners of baseplate or next to column web (see Appendix and HD bolt manufacturers' catalogues).
3. Maximum allowable bearing strength $= 0.4f_{cu}$ (code 4.13.1)
   where $f_{cu} = 28$ day characteristic cube strength of the concrete.
4. Actual bearing pressure to be less than or equal to $0.4f_{cu}$.
5. Determine baseplate thickness t: (code 4.13.2.2)
   For I, H, channel, box or RHS columns
   $t = 2.5 \, w/p_{yp} \, (a^2 - 0.3b^2)^{\frac{1}{2}}$ but not less than the thickness of the flange of the supported column.
   $w =$ pressure in N/mm^2 on underside of plate, assuming a uniform distribution.

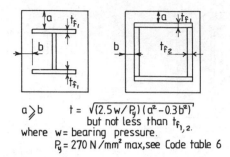

$a \geqslant b$ $t = \sqrt{(2.5\,w/P_y)\,(a^2 - 0.3b^2)}$
          but not less than $t_{f_{1,2}}$.
where w= bearing pressure.
     P_y= 270 N/mm² max, see Code table 6

Fig. 3.11 Baseplates subjected to concentric forces (code 4.13.2.2). (Courtesy of the British Standards Institution.)

a = larger plate projection from column.
b = smaller plate projection from column.
p_{yp} = design strength of plate, but not greater than 270 N/mm².
See Fig. 3.11. Note Table 3.2 gives nominal bearing strength of different grades of concrete.
Weld check:
6. Calculate length of weld to resist axial load
   i.e. $4B + 2D + 2D - 2t_w$
   B = column width,
   D = column depth,
   t_w = web thickness.
7. Select weld size.
8. Check shear stress on weld.
9. Vector sum of all the stresses carried by the weld must not exceed p_w, the design strength.   (code 6.6.5.5)

Table 3.2 Nominal bearing strengths for concrete.

Conc. grade	15	20	25	30	40	50	60
Nom. brg. N/mm²	6	8	10	12	16	20	24

Bolt check   (code 6.7)
10. Check maximum allowable simultaneous shear (and tension, if any), on the holding down bolts, from manufacturer's tables (see also Appendix); or if threaded bars used, by criteria similar to that specified for ordinary bolts.

3.5.2 Procedure for determining a baseplate pressure diagram due to axial load and moment – fixed connection

1. Obtain values of moment and axial load (ultimate values).
2. Refer to Fig. 3.8 and assume values for x, l_a, and l.
3. $e = M/P$.
4. If $e > \dfrac{d_1}{2}$ and $l/6$ continue calculation based on Fig. 3.12 and proceed to step 5.

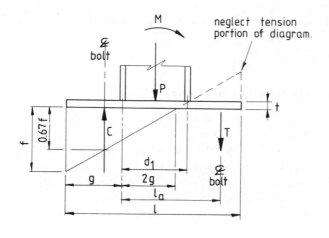

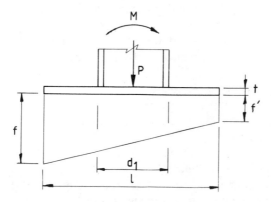

Fig. 3.12 Fixed column baseplate with axial load and moment.

If $e < \dfrac{d_1}{2}$ but $> l/6$ ⎤ proceed from step 10 and base

If $e < l/6$ ⎦ calculation on Fig. 3.12.

5. Assume a location for C (bolt centreline).

6. $T = P\left(e - \dfrac{d_1}{2}\right)\bigg/ l_a \qquad T = P(e - d_1/2)/l_a$

7. $C = T + P$

8. $f = 2C/3gb$ where $f \leqslant 0.4 f_{cu}$, where $b =$ width of baseplate

9. End of calculation

10. Calculation continued from step 4. Refer to Fig. 3.12.

11. $f = \dfrac{P}{A_{plt}} + \dfrac{M}{Z_{plt}}$, where $Z_{plt} =$ elastic modulus of baseplate.

$$= bl^2/6$$

NOTE: $f \leqslant 0.4$; f_{cu} and A_{plt} = area of baseplate.

12. $f = \dfrac{P}{A_{plt}} - \dfrac{M}{Z_{plt}}$

3.5.3 Procedure for design of a stiffened baseplate – fixed connection

1. Obtain axial loads, shear and moment at base (ultimate).
2. Decide on number and type of HD Bolts to resist the most unfavourable combination of shear, axial load and moment. (Shear to be carried equally by all bolts.) Position bolts such that the lever arm between bolt groups is sufficient to ensure that the moment of resistance is greater than the applied moment. Sketch preliminary base configuration.
3. Draw pressure diagram to underside of baseplate (see Fig. 3.12) (Elastic pressure).
4. Using maximum pressure on diagram, check that concrete bearing capacity $(0.4 f_{cu})$ is not exceeded.
5. If the bearing pressure is not uniform, determine the maximum bending moment in the baseplate which shall not exceed $1.2\,p_{yp}Z$.   (code 4.13.2.3)
   $p_{yp} \leqslant 270\,\text{N/mm}^2$
   Z = elastic modulus of baseplate
6. Having determined details of baseplate, bolts and location of stiffeners, check bending capacity of stiffeners as cantilevered from the column. The moment should not exceed $p_{yg}Z$.   (code 4.13.2.4)
   p_{yg} = design strength of stiffener $\leqslant 270\,\text{N/mm}^2$
   Z = elastic modulus of stiffener.
7. Check weld on stiffeners for shear and bending moment. (Use nominal weld for stiffeners and column to baseplate, if possible.)
8. Check that shear and tension on bolts is within the allowable.
9. Sketch the connection.

3.5.4 Column baseplate design examples

Example 3.1. Pinned column baseplate (see Section 3.5.1)

Design a pinned base with the connection data given below:

Column	$203 \times 203\text{UC}46$
Ultimate axial load	200 kN compression
Ultimate shear load	80 kN (acting in the web axis)
Concrete grade 20	$f_{cu} = 20\,\text{N/mm}^2$
Holding down bolts	grade 43 steel.

SOLUTION
Try baseplate 220 mm square
Bearing strength for grade 20 concrete = 8 N/mm² (Fig. 3.7)

Actual bearing stress $= 200 \times 10^3/220 \times 220 = 4.13\,\text{N/mm}^2$

$< 8\,\text{N/mm}^2$

Baseplate thickness t

$$t = [2.5w/p_{yp}(a^2 - 0.3b^2)]^{\frac{1}{2}} \qquad \text{(code 4.13.2.2)}$$

$a = b = 10\,\text{mm (approx.)}$

$\therefore t = [2.5 \times 4.13/270(10^2 - 0.3 \times 10^2)]^{\frac{1}{2}}$

$= 1.7\,\text{mm}.$

Base thickness should not be less than the column flange thickness, which in this case equals 11 mm. For practical reasons use 15 mm thick baseplate.

Use baseplate $220 \times 220 \times 15\,\text{mm}$

Bolts

Try 2 No. M24 HD bolts

Max. allowable shear on 2 No. M24 HD bolts (grade 4.6)

$2p_s A_s = 2 \times 160 \times 353/10^3$

$= 113\,\text{kN} > 80\,\text{kN shear}$

Use 2 No. M24 HD bolts (grade 4.6)

Weld (stanchion to baseplate)

Ultimate axial load $= 200\,\text{kN}$

Ultimate shear load acting on web only

Axial load on weld $= 200\,\text{kN/column perimeter}$

$= 200/1187$

$= 0.17\,\text{kN/mm}$

Shear load on web weld $= 80/2 \times \text{distance between root fillets}$

$= 80/2 \times 160.8$

$= 0.25\,\text{kN/mm}$

Max. resultant load on weld $= \sqrt{(0.17^2 + 3 \times 0.25^2)}$

$= 0.47\,\text{kN/mm}.$

Use 6 mm profile weld to column/baseplate (strength $= 0.90$ kN/mm – Fig. 3.8).

Example 3.2. Fixed column baseplate (see Section 3.5.2)

Design a fixed base with the connection data given below:

- Column $305 \times 305\text{UC}97$
- Ultimate axial load $=$ 100 kN (compression only)
- Ultimate shear load $=$ 34 kN (at $90°$ to the xx axis)
- Ultimate moment $=$ 83.9 kN m (about the xx axis)
- Concrete grade $=$ $20\,\text{N/mm}^2$
- Holding down bolts grade 43 steel

SOLUTION

Try baseplate 550 square with bolts 75 from edge (Fig. 3.12)

$l = 550\,\text{mm}$

$g = (550 - 307.8)/2 + 15.4/2 = 128.8\,\text{mm}$

(assume centre of compression is on centreline of column flange.)

$e = M/P = 83.9 \times 10^3/100 = 839$ mm.

$l/6 = 550/6 = 91.67 < 839$

$d_1 = 307.8 - 15.4 = 292.4$

$d_1/2 = 292.4/2 = 146.2 < 839$

$l_a = 550/2 + 292.4/2 - 75 = 346.2$ mm.

$T = P(e - d_1/2)/l_a = 100(839 - 146.2)/346.2$
$= 200$ kN

$C = T + P = 200 + 100 = 300$ kN

Overall baseplate size

Bearing capacity $f = 2C/3gb$

$f = 2 \times 300 \times 10^3/3 \times 128.8 \times 550 = 2.85$ N/mm^2 < 8 N/mm^2

Bearing adequate.

Bolts

Use four bolts.

$T = 100$ kN per bolt

Shear per bolt $= 34/4 = 8.5$ kN.

Use M30 bolts (Fig. 3.1).

Weld

Shear in direction of web $= 34$ kN

Length of web $= 307.8 - 2 \times 15.4 = 277$ mm.

Shear per mm of web weld $= 34 \times 10^3/277 \times 2 = 0.06$ N/mm

Axial load on column flange $= 300$ kN

Length of weld $= 1800$ mm.

Axial load per mm weld $= \dfrac{300}{1800} = 0.16$ kN/mm

Vector sum of load on weld $= (0.06^2 + 0.16^2)^{\frac{1}{2}} = 0.17 < 0.90$ kN/mm (Fig. 3.8).

Use 6 mm perimeter weld to baseplate/column

Baseplate.thickness

Max. cantilever moment in baseplate (due to bolt tension):

$M = T \times$ (bolt centreline to flange edge)

$= 200$ kN $\times 4.61$ cm.

$= 922$ kN cm per column width of 305 mm.

$t = (922 \times 6/27 \text{ kN/cm}^2 \times 30.5)^{\frac{1}{2}} = 2.6$ cm

Use 30 mm. thick baseplate (> 26 mm.)

3.6 TENSION CONNECTIONS (see Fig. 3.13)

Bolted tension connections should generally be designed for plate bending (between the bolt axis and loading axis). Inbuilt factors of safety for the allowable bolt tension more than compensate for any prying action that may occur (code 6.3.6.2). The concept of prying action is illustrated in Fig. 3.14. A minimum value for prying action on the bolt may be assumed at not less than 10% of the bolt axial load. Alternatively, in cases where the designer considers the

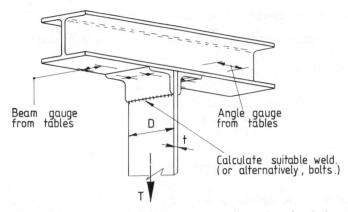

Fig. 3.13 Hanger connection arrangement dimensions for design.

connection warrants closer study he may depart from the code assumption that prying is allowed for, and conservatively assume a 30% prying force, or some lesser value using the method of calculation outlined in BS5400: Part 3, Clause 14.3.6. However, prying is particularly sensitive to plate fit-up and therefore can only be approximately estimated.

For welded tension connections across a beam flange (at 90° to the web) the load is carried by the web and plate over a 'stiff' length (45° dispersion from web), and weld should be *kept within this zone*, or the flange suitably stiffened.

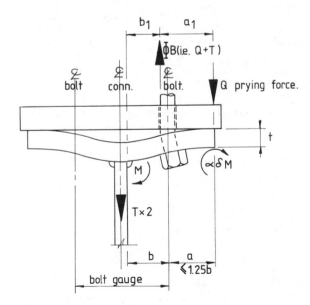

Fig. 3.14 Bolt prying action (after Struik and deBach).

3.6.1 Procedure for design of bolted connections

(1) Bolt tension T = ultimate tensile force on bolt group divided by the number of bolts in group, securing plate at 90° to member axis.

(2) Bolt shear S = ultimate shear force on the connection divided by the total number of bolts in the connection plane

(3) Select a suitable bolt size and grade (from Fig. 3.1).

(4) Assume 45° load spread on the bolted plate, from the bolt centreline to the load centreline.

(5) Design plate for the width obtained by 45° load dispersion and for plastic action, with an elastic modulus of $bd^2/4$.

2.6.2 Tension connection design example

Example 3.3

A tension connection is required between a hanger plate and a universal beam, as illustrated in Fig. 3.13. The hanger plate is attached to the beam by 2 angles with a total of 4 bolts. The beam section is 305×165 UB 54. The hanger plate width and thickness are 165 mm × 10 mm, and the tension in the hanger is 60 kN (ultimate load).

Design a suitable connection; using grade 43 steel and grade 4.6 bolts.

SOLUTION
Try 2 No. angles $100 \times 100 \times 12$ with 4 No. M16 bolts connecting to the beam and fillet welds connecting to the hanger plate.

$$T = \frac{\text{Total ultimate tension}}{\text{No. of bolts}} = \frac{60\,\text{kN}}{4} = 15\,\text{kN} < 30.6\,\text{kN (Fig. 3.1)}$$

Shear on connection = 0 (no shear on bolts)

Use 4 NO. M16 bolts (grade 4.6).

Check plate bending on angles
Bolt backmark = 55 mm
Distance from bolt centreline to inside edge of leg = $55 - 12 = 43$ mm
Total length of angle used = beam bolt gauge + 2 × edge dist. = $90 + 2 \times 30 = 150$ mm = 75 mm per bolt.
(load dispersion in angles at 45° or less cannot exceed 150 mm)

$$f = \frac{M}{S} = \frac{15\,\text{kN} \times 0.043\,\text{m} \times 10^6 \times 4}{75 \times 12^2 \times 2} \quad \text{(where } S = bd^2/4 \text{ and } M = WL/2\text{)}$$

$$= 119.5\,\text{N/mm}^2 < 275$$

Use 2 NO. $100 \times 100 \times 12$ angles

Check plate bending. Universal beam
Load dispersion from bolt at 45° to web
Bolt gauge = 90 mm; bolt load = 55 kN
Width of load dispersion = bolt centres on angles plus dispersion at 45° to web
$$= 2 \times 55\,\text{mm} + 10\,\text{mm} + 2(45 - \tfrac{1}{2}\text{web}), \text{ (web} = 7.7\,\text{mm)}$$
$$= 202\,\text{mm (i.e. 101 mm per bolt)}$$

$p_b = 15 \times 0.041 \times 10^6 \times 4/(101 \times 13.7^2 \times 2) = 65 \, \text{N/mm}^2$
Note: check combined stresses on beam do not exceed allowable.
Check weld on hanger plate/angles
Total tensile force $= 60 \, \text{kN}$
Continuous fillet weld to connect plate to angles on both sides. Try 6 mm fillet weld.
Length of 6 mm fillet weld required to resist vertical shear
    $= 60/2 \times 0.90 \, \text{kN}$ (Fig. 3.8)
    $= 27 \, \text{mm}$ each side.

Use 6 mm continuous fillet weld (both sides) to connect hanger plate to angles.

3.7 SPLICE CONNECTIONS

Structural members are spliced basically for three reasons:

(1) Ease of fabrication
(2) Ease of transportation
(3) Ease of erection.

One or a combination of these factors dictate the necessity for a splice. In

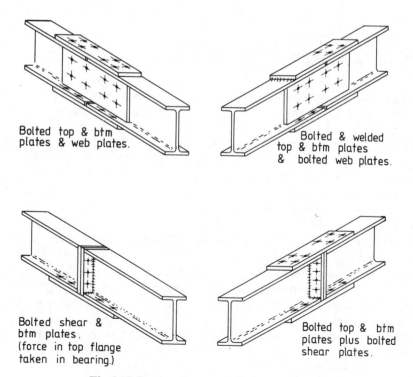

Fig. 3.15 Beam splices – typical arrangement.

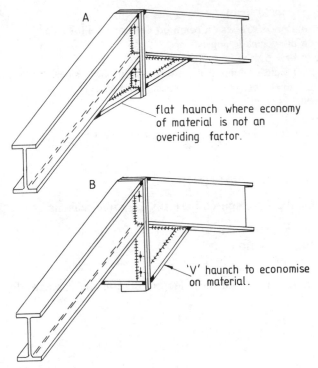

flat haunch where economy
of material is not an
overiding factor.

'V' haunch to economise
on material.

Fig. 3.16 Typical portal frame ridge connection.

addition, rolled steel sections are normally supplied in stock lengths of 12 m (in the United Kingdom) but can be supplied in longer lengths by special request.

Three types of splices are illustrated (see Figs. 3.15, 3.16 and 3.17), beam splices, column splices and a 'special' beam splice – the portal frame ridge connection. The main criterion of design for splice connections is that of selecting the cheapest satisfactory solution bearing in mind the work involved in fabrication and the difficulties of erecting the final product.

The design of splices consists basically of a check on the carrying capacity of the bolts and the plys, which is covered in Chapter 4. Tension can be carried by flange plates or a web plate, but if both flange and web plates are used the tension should be assumed to be shared between flanges and web in proportion to the respective areas joined. Column splices should be designed for the extremes of maximum uplift and maximum compression because it is not usual to end machine columns.

The portal frame ridge connection is designed in the same way as a beam to column moment connection (see Chapter 4). When splice flange plates are long (distance between first and last bolts on side of a splice > 500 mm), reference

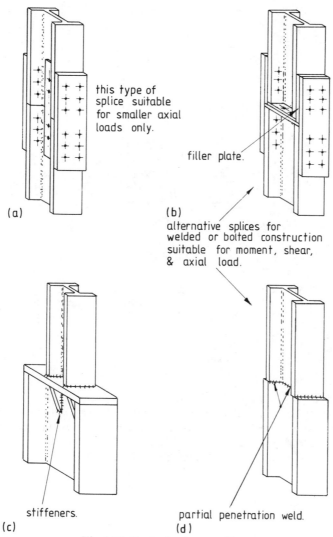

this type of
splice suitable
for smaller axial
loads only.

(a)

filler plate.

(b)

alternative splices for
welded or bolted construction
suitable for moment, shear,
& axial load.

stiffeners.

(c)

partial penetration weld.

(d)

Fig. 3.17 Typical column splices.

should be made to the code 6.3.4 for a reduction factor on the allowable shear
capacity of the fasteners due to plate extension.

3.7.1 Column splice design example

Example 3.4

Two columns are to be spliced for full fixity. Design a suitable connection given the
following ultimate loads and moments:

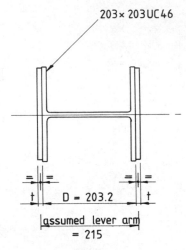

203 × 203 UC 46

$D = 203.2$

assumed lever arm = 215

Fig. 3.18

Column size (above and below splice) 203 × 203 UC 46
Axial load at splice   200 kN (tension) or 500 kN (compression)
M_{xx} at splice   50 kNm max.
M_{yy} at splice   0 kNm
Shear xx axis   80 kN
Shear yy axis   0 kN

SOLUTION
Use arrangement as shown in Fig. 3.17(b).
Assume axial load and M_{xx} taken on flange plates, and shear$_{xx}$ taken on end plates.
Flange plates (to carry axial load, moment + moment caused by the eccentric loading of the flange plates) (Fig. 3.18):
Axial load per flange plate $= 200/2 = 100$ kN tension
                or $500/2 = 250$ kN compression (worst case)
Load on flange plate due to $M_{xx} = 50/0.215 = 232.6$ kN
Column p_c $= 200$ N/mm^2
Additional $M = S(p_y - p_c)$
             $= 497.4 \times 10^{-3}\ (275 - 200)$
             $= 37.3$ kNm
Load on flange plate due to additional moment $= 37.3/0.215 = 173.4$ kN
Total load on flange plate $= 250 + 232.6 + 173.4 = 656$ kN
$t = \text{load}/(\text{plate width} \times p_y) = 656 \times 10^3/(200 \times 275) = 11.9$ mm.
Use 12 mm flange plates

Bolts to flange plates

Try 6 No. bolts each side of splice
Shear per bolt $= 656/6 = 109.3$ kN < 132.4 (Fig. 3.1)
Use 6 No. M24 grade 8.8 bolts (each side of splice)

End plates (to carry shear)
Shear load on end plate = 80 kN.
Plate thickness governed mainly by bolt bearing capacity
Assume nominal thickness of 10 mm. and check suitability as part of bolt calculation.

Use 12 mm. thick end plates

Bolts to end plates

Try 4 no. bolts (2 each side of column web)
Shear per bolt = 80/4 = 20 kN.

Use 4 No. M16 grade 8.8 bolts

(2 each side of column web, symmetrically arranged)

Weld to end plates (profile weld to column perimeter, shear in xx axis resisted by weld

between the column root fillets 'd').
Shear on weld between root fillets = $80 \times 10^3/2 \times 160.8$
$$= 248.7 \, N/mm$$
$$= 0.25 \, kN/mm.$$

Use 5 mm. fillet weld to column profile (0.76 kN/mm capacity)

3.8 BRACING CONNECTIONS

Generally bracing is designed to act in simple compression or tension. Bracing normally consists of a single member with two or more bolts at each end connecting the member to a plate. Since the member is joined to a structural system by a minimum of two bolts at each end, it is considered to be fixed in position and partially fixed in direction at both ends, with a slenderness ratio of $0.85 \, L/r_{min}$ but not less than $(0.7 \, L/r) + 30$. These slenderness ratios and those for other bracing forms are given in code 4.7.10.

When single angles are used to form bracing, axial load in the angle passes (in theory) through the centroid of the member cross-section. The centroid and bolt backmark do not coincide, however, creating a moment on the connection. With smaller connections this can safely be ignored, but with larger connections the moments should be calculated and designed for – see code 4.7.6 – in addition to the axial forces on the joint.

Several typical small bracing connections are illustrated (Figs 3.19 and 3.20) but there is, of course, no limit to the variation of bracing connections, bearing in mind the limits of practical design and the appearance of the finished connection. Calculation for this type of connection is generally limited to a check of bolt shear and bearing values, and weld shear value.

With larger connections such as the heavy gusset plates that occur on bridges, it is possible to make an approximate check of the plate suitability by selecting various planes passing through critical points and checking for allowable stress on the basis of $P/A \pm M/Z$ in the traditional (elastic) fashion. More complicated analyses using computer finite element methods have been found to be not

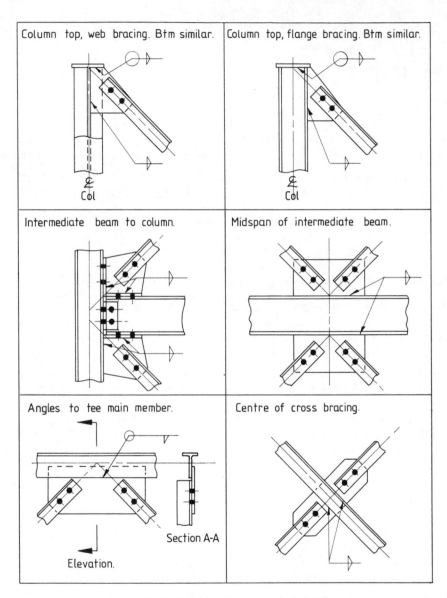

Fig. 3.19 Vertical bracing – typical details.

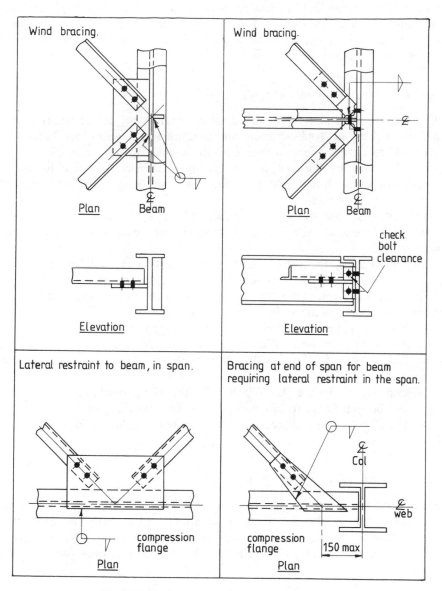

Fig. 3.20 Horizontal bracing – typical details.

strictly applicable and, in the absence of research data on the subject, the authors recommend calculation based on the 'traditional' method. Where the joint concerned will rotate out of plane as the structure deflects under applied load, the plate carries the rotation as a uniform moment which should be included in the design. A local check on elastic shear should be carried out, as indicated in Fig. 2.7, Chapter 2.

3.9 BEAM TO BEAM CONNECTIONS

When secondary beams frame into primary beams, they are generally placed at the same 'top of steel' elevation, and notched as shown in Fig. 3.21a. It is advisable to have a nominal radius at the corner of the notch. The bolted connection can be checked as an eccentric connection (see Chapter 4, Section 4.6) although a simple check for shear and bearing on the bolts and plate is usually sufficient.

The notched beam should also be checked for shear and moment capacity at the end of the notch, although it is exceptional for the capacity to be exceeded in most cases.

Maximum elastic shear occurs at the neutral axis of the notched shape, and may be found using the formula for elastic shear quoted in Chapter 2, Section 2.2.4. Maximum bending stress occurs at the top of the single notched shape, and should be determined using the elastic modulus (Z).

A distance (w) of unsupported web has been suggested by Pask [4] to ensure that the beam web has sufficient stability to develop allowable stresses (Fig. 3.21(a)). For definition of a flexible connection refer to the introduction of Chapter 4.

Welded connections are shown in Fig. 3.21(b). The one-sided connection has a 'sniped' bottom flange in order to prevent local buckling of the main beam web. The double sided connection is not sniped, to allow continuity of the secondary beam.

3.10 TUBULAR CONNECTIONS

Tubular members have particular advantages not possessed by conventional sections; these make their use in certain locations commercially viable and even preferable in some instances, such as in areas of high torsion. With the advent of lattice space frames for roofs, the tube has found widespread adoption for reasons of

(1) High strength to weight ratio
(2) Good torsional properties
(3) Smart appearance.

Where offshore structures are concerned, the tube is often selected for the

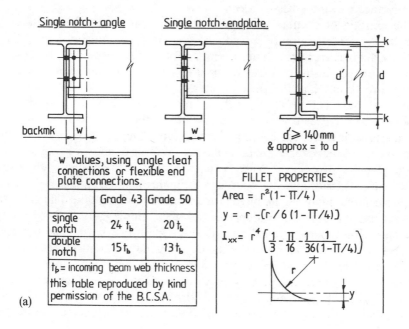

Single notch + angle   Single notch + endplate.

backmk | w

$d' \geqslant 140$ mm
& approx = to d

w values, using angle cleat connections or flexible end plate connections.		
	Grade 43	Grade 50
single notch	$24\,t_b$	$20\,t_b$
double notch	$15\,t_b$	$13\,t_b$

t_b = incoming beam web thickness

this table reproduced by kind permission of the B.C.S.A.

(a)

FILLET PROPERTIES

$$\text{Area} = r^2(1 - \pi/4)$$

$$y = r - (r/6\,(1 - \pi/4))$$

$$I_{xx} = r^4\left(\frac{1}{3} - \frac{\pi}{16} - \frac{1}{36(1-\pi/4)}\right)$$

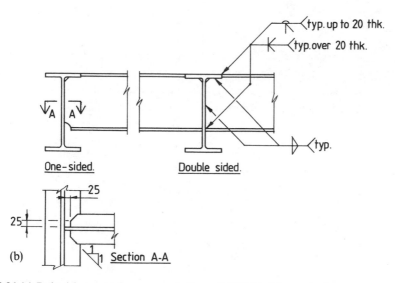

⟨typ. up to 20 thk.
⟨typ. over 20 thk.

One-sided.   Double sided.

⟨typ.

(b)   Section A-A

Fig. 3.21 (a) Bolted beam to beam connections. (b) Welded beam to beam connections.

above reasons, and in addition because of the smaller effective area that it offers as resistance to wind gusts or ocean waves.

The jointing of circular tubes poses a particular problem because of the complexity of intersection of curved surfaces. However, fully automatic flame-cutting machines are now available to cut appropriate tube saddles. The machine is preset for the outside diameter of the intersected tube, the inside diameter of the tube to be cut, and the angle of intersection and, as the machine cuts the tube, it automatically puts a bevel on the edge for welding purposes. It is preferable to run tube bracing to node points on the main member (Fig. 3.22a) but not if this involves cutting tubes for two saddle shapes (Fig. 3.22b). In order to avoid double cuts, tubes are often moved off-node as shown Fig. 3.22c either in one plane or two planes. When bracing is offset in this fashion, full account must be taken of the moments induced in the main member. On major structures, where the forces involved are often large, connections may take different forms, such as tubes running into a sphere, or tubes running into a locally larger diameter main member (called a 'can') (Fig. 3.23) or even passing through a member (Fig. 3.22d). In addition to the need for individual members and welds to be properly designed to withstand the axial loads and moments (and hydrostatic pressure if any), they must be designed against punching shear where a smaller brace member exerts a force normal to the wall of the larger main member. Foundation connections for tubular columns may be designed

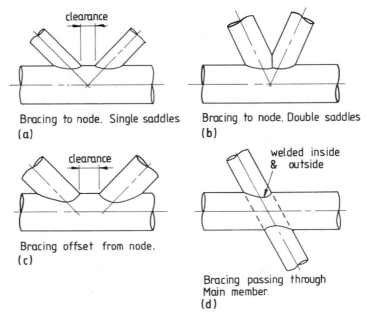

Bracing to node. Single saddles
(a)

Bracing to node. Double saddles
(b)

Bracing offset from node.
(c)

Bracing passing through
Main member.
(d)

Fig. 3.22

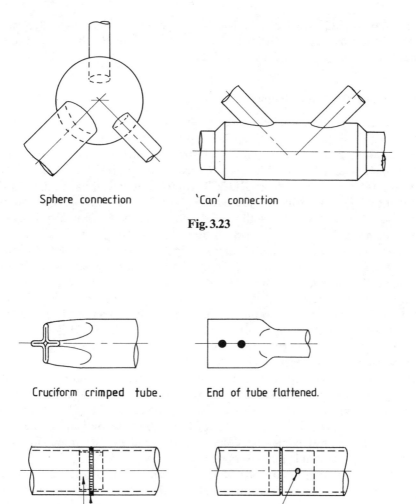

Sphere connection   'Can' connection

Fig. 3.23

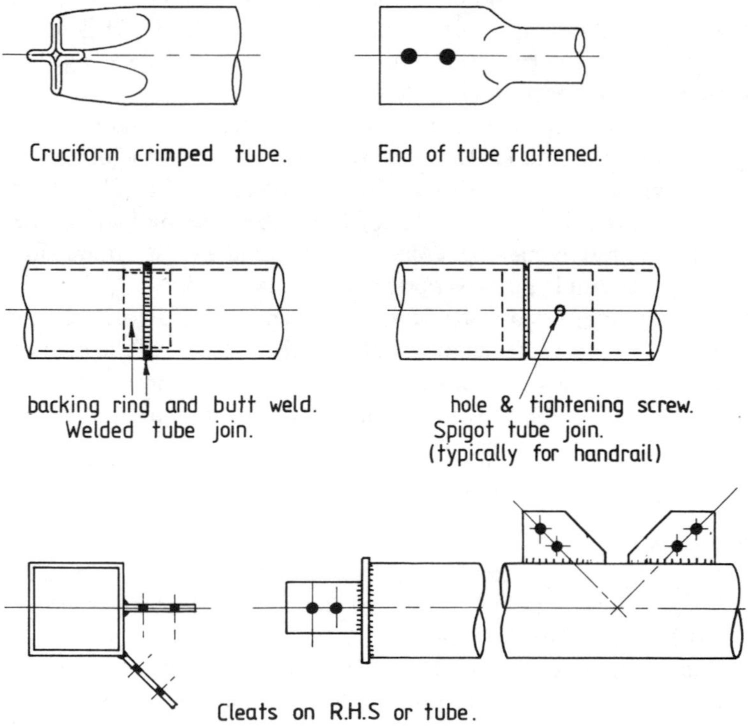

Cruciform crimped tube.   End of tube flattened.

backing ring and butt weld.
Welded tube join.

hole & tightening screw.
Spigot tube join.
(typically for handrail)

Cleats on R.H.S or tube.

Fig. 3.24 Tube connections for minor structures.

in a manner similar to that adopted for any other column baseplate, bearing in mind the usual design parameters. A selection of small tube connections for small structures is illustrated in Fig. 3.24.

3.10.1 Design of welded tubular joints

Design values for fully welded tubular joints (with tube centrelines in one plane) have been produced with the financial help of the Commission of the European Communities and the participation of the International Committee for the Development and Study of Tubular Construction (CIDECT). The following text and tables are based on a CIDECT monograph which has been published by British Steel, and which is reproduced here by their kind permission, with the assistance of Messrs T. Giddings, J. Hallum and N. F. Yeomans. The lattice girder example following was also produced by British Steel.

The recommendations given in Tables 3.3 to 3.6 are valid for steels to Euronorm 25–72 of Fe360, Fe430 or Fe510. Equivalent steels must have chemical and mechanical properties as noted in Euronorm 25, with minimum yield strengths as noted in Fig. 3.25. British Steel sections are available in grades 43 and 50 (to BS4360) which correspond to Euronorm 25 Fe430 and Fe510 respectively. Special through-thickness (Z) steel (see Chapter 1) should be specified for chords whose flange or wall thickness exceeds 25 mm. Fig. 3.26 indicates properties of hollow sections that must be checked at the delivery stage.

Basic joint types are summarized in Fig. 3.27. Joints K, N, and KT with branch forces all acting in the same sense (i.e. all struts or all ties) should be checked as being X joints. Joints are assumed to be *pinned* for the sake of simplicity, and even though with continuous chord members, secondary moments will be set up in the bracing members, the tabulated checks allow the load analysis to be based on a simple pinned framework assumption.

Nomenclature for the joint calculations is set out below in an abbreviated form. Subscripts adjacent to symbols indicate the location of the member under consideration; for instance 'o' indicates chord, 'i' indicates brace member. When $i = 1$ compression is indicated, and when $i = 2$ tension is indicated.

CI Joints: CHS Bracing, I chord
RI Joints: RHS Bracing, I chord

e : eccentricity of joint
d_o: outer diameter of a circular chord member
b_o: external width of a square or rectangular chord member
t_o : wall thickness of the chord member (CHS or RHS)
d_i: outer diameter of a circular bracing member (accepted notation; $i = 1$ for the compression bracing and $i = 2$ for the tensile bracing member)
b_i: external width of a square or rectangular bracing member
t_i : wall thickness of a bracing member

Table 3.3 X, T, Y joints with RHS chords

Application limits:
$$\frac{b_i}{t_i} \text{ and } \frac{h_i}{t_i} \leq 35 \text{ in tension and } 0.9\sqrt{\frac{E}{f_{yi}}} \text{ in compression}$$
$$\frac{d_i}{t_i} \leq 50 \text{ in tension and } 1.2\sqrt{\frac{E}{f_{yi}}} \text{ in compression}$$
$$0.4 \leq \beta \leq 1.0$$
$$0.5 \leq \frac{h_i}{b_i} \leq 2.0$$
$$30° \leq \theta \leq 90°$$

Type	Formula	Failure criterion	Ultimate strength	Definition of parameters	γ_m
RHS bracing members $\beta = 1.0*$	1.1	Plastification or local buckling of the walls	$N_k = \dfrac{\sigma_k t_o}{\sin\theta_i}\left(\dfrac{2h_i}{\sin\theta_i} + 10t_o\right)$	– in tension: $\sigma_k = f_{yo}$ – in compression: $\sigma_k = \sigma_{uN}$ calculated by means of the buckling curves with the following slendernesses: T and Y joints: $\lambda = \dfrac{2.63\left(\dfrac{h_o}{t_o}-2\right)}{\sqrt{\sin\theta_1}}$ X joints: $\lambda = \dfrac{3.46\left(\dfrac{h_o}{t_o}-2\right)}{\sqrt{\sin\theta_1}}$	1.0 in tension 1.2 in buckling
RHS bracing members $\beta = 0.85*$	1.2	Plastification of the chord	$N_k = \dfrac{2f_{yo}t_o^2}{(1-\beta)\sin\theta_i}\left[\dfrac{h_i}{b_o\sin\theta_i} + 2(1-\beta)^{0.5}\right]$		1.0
RHS bracing members $0.85 < \beta \leq 1.0$	1.3	Effective width of the bracing member	$N_k = f_{yi}t_i(2h_i - 4t_i + 2b_e)$	$b_e = \dfrac{Ct_o}{b_o}\left(\dfrac{f_{yo}t_o}{f_{yi}t_i}\right)b_i \leq b_i$ with $C = 13.5$ for Fe 360 and Fe 430 steels $C = 11.5$ for Fe 510 steel and $\left(\dfrac{f_{yo}t_o}{f_{yi}t_i}\right) \geq 1.0$	1.25 on b_e only
RHS bracing member $0.85 < \beta \leq \dfrac{b_o - 2t_o}{b_o}$	1.4	Shear	$N_k = \dfrac{1.15f_{yo}}{\sin\theta_i}\left(\dfrac{h_i}{\sin\theta_i} + b_{ep}\right)t_o$	$b_{ep} = \dfrac{Ct_o}{b_o}b_i \leq b_i$ with C as above	1.25 on b_{ep} only
CHS bracing members $\beta \leq 0.8$	1.5	Plastification of the chord	$N_k = \dfrac{1.57f_{yo}t_o^2}{(1-\beta)\sin\theta_i}\left[\dfrac{d_i}{b_o\sin\theta_i} + 2(1-\beta)^{0.5}\right]$		1.0

*For β between 0.85 and 1, use linear interpolations between formulae 1.1 and 1.2.

Table 3.4 K and N joints with RHS chords, gap and overlap joints

Type	Formula	Failure criterion	Ultimate strength	Application limits: $0.4 \leqslant \beta \leqslant 1.0$ $0.5 \leqslant \dfrac{h_i}{b_i} \leqslant 1.0$ $30° \leqslant \theta \leqslant 90°$	$\dfrac{b_i}{t_i}$ and $\dfrac{h_i}{t_i} \leqslant 35$ in tension and $0.9\sqrt{\dfrac{E}{f_{yi}}}$ in compression $\dfrac{d_i}{t_i} \leqslant 50$ in tension and $1.2\sqrt{\dfrac{E}{f_{yi}}}$ in compression		

Type	Formula	Failure criterion	Ultimate strength	Definition of parameters	γ_m
RHS bracing members, gap joints	2.1	Plastification of the chord	$N_{1k} = \dfrac{6.93 f_{yo} t_o^2}{\sin\theta_1} \beta^* \left(\dfrac{b_o}{t_o}\right)^{0.5} \eta$ $N_{2k} = N_{1k}\dfrac{\sin\theta_1}{\sin\theta_2}$	$\beta^* = \dfrac{b_1 + b_2 + h_1 + h_2}{4b_o} \quad \beta^{**} = \dfrac{b_1 + b_2}{2b_o}$ $\eta = 1.0$ when the chord is in tension $\eta = 1.3 - \dfrac{0.4}{\beta^{**}}\left|\dfrac{\sigma_o}{f_{yo}}\right| \leqslant 1.0$ when the chord is in compression ($\sigma_o \leqslant 0$ in the chord)	1.1
$0.5(1 - \beta^{**}) \leqslant \dfrac{g}{b_o}$ $\dfrac{g}{b_o} \leqslant 1.5(1 - \beta^{**})$	2.2	Shear	$N_k = \dfrac{f_{yo}}{\sqrt{3}\sin\theta_i} A_Q$	$A_Q = 2h_o t_o + \alpha b_o t_o$ with $\alpha = \sqrt{\dfrac{1}{1 + 1.25\left(\dfrac{g}{t_o}\right)^2}}$	1.0
	2.3	Shear	$N_k = (A_o - A_Q)f_{yo} + A_Q f_{yo}\sqrt{1 - \left(\dfrac{Q}{Q_p}\right)^2}$	$Q = (N_i \sin\theta_i)_{\max} \quad Q_p = \dfrac{A_Q f_{yo}}{\sqrt{3}}$	1.0

2.4	Effective width of bracing member	$N_k = f_{yi}t_i(2h_i - 4t_i + b_i + b_e)$	$b_e = \dfrac{Ct_o}{b_o}\left(\dfrac{f_{yo}t_o}{f_{yi}t_i}\right)b_i \leq b_i$ with $\left(\dfrac{f_{yo}t_o}{f_{yi}t_i}\right) \geq 1.0$ with $C = 13.5$ for Fe 360 and Fe 430 steels $C = 11.5$ for Fe 510 steel	1.25 for the b_e term only
2.5	Shear	$N_k = \dfrac{f_{yo}t_o}{\sqrt{3}\sin\theta_i}\left(\dfrac{2h_i}{\sin\theta_i} + b_i + b_{ep}\right)$	$b_{ep} = \dfrac{Ct_o}{b_o}b_i \leq b_i$ with C as above	1.25 for the b_{ep} term only
2.6	Effective width of bracing member RHS bracing members overlap joints $30\% \leq u \leq 100\%$	$N_k = f_{yi}t_i(2h_i - 4t_i + b_e + b_{eu})$ b_{eu} refers to the "covering" bracing member	$b_e = \dfrac{Ct_o}{b_o}\left(\dfrac{f_{yo}t_o}{f_{yi}t_i}\right)b_i \leq b_i$ with $\left(\dfrac{f_{yo}t_o}{f_{yi}t_i}\right) \geq 1.0$ $b_{eu} = \dfrac{Ct_i}{b_i}\left(\dfrac{f_{yiu}t_{iu}}{f_{yi}t_i}\right)b_i \leq b_i$ with $\left(\dfrac{f_{yiu}t_{iu}}{f_{yi}t_i}\right) \geq 1.0$ and C as per formula 2.4	1.25 for the b_e and b_{eu} terms only
2.7	Plastification of the chord RHS bracing members gap joints $\beta^{**} \leq 0.8$ and $0.5(1 - \beta^{**}) \leq \dfrac{g}{b_o}$ $\dfrac{g}{b_o} \leq 1.5(1 - \beta^{**})$	$N_k = \dfrac{5.43 f_{yo}t_o^2}{\sin\theta_1}\beta^{**}\left(\dfrac{b_o}{t_o}\right)^{0.5}\eta$	$\beta^{**} = \dfrac{d_1 + d_2}{2b_o}$ and η as per formula 2.1	1.1

Table 3.5 X, T, Y, K and N joints between CHS

Type	Formula	Failure criterion	Ultimate strength	Application limits: $0.25 \leq \beta \leq 1.0$ $30° \leq \theta \leq 90°$ $\;$ Definition of parameters	$d_i/t_i \leq 50$ in tension and $1.2\sqrt{\dfrac{E}{f_{yi}}}$ in compression $\quad$ γ_m
X joints	3.1	Plastification of the chord	$N_k = \dfrac{5.84 f_{yo} t_o^2}{\sin\theta_i}\left(\dfrac{1}{1-0.81\beta}\right)\eta$	$\eta = 1.0$ when $\dfrac{\sigma_o}{f_{yi}} \geq -0.4$ $\eta = 1.2 - 0.5\left\|\dfrac{\sigma_o}{f_{yi}}\right\|$ when $\dfrac{\sigma_o}{f_{yi}} < -0.4$ giving σ_o the minus sign in compression	1.1
T and Y joints	3.2	Plastification of the chord	$N_k = \dfrac{2.8 f_{yo} t_o^2}{\sin\theta_i}(1+4.94\beta^2)\left(\dfrac{d_o}{t_o}\right)^{0.2}\eta$	η as above	1.1
K and N joints (gap or overlap joints)	3.3	Plastification of the chord	$N_{1k} = \dfrac{1.71 f_{yo} t_o^2}{\sin\theta_1}(1+4.67\beta)\eta'\gamma$	$\eta' = 1.0$ when $\dfrac{N_{opd}}{A_o f_{yo}} \geq 0$ $\eta' = 1 + 0.3\dfrac{N_{opd}}{A_o f_{yo}} - 0.3\left(\dfrac{N_{opd}}{A_o f_{yo}}\right)^2$ when $\dfrac{N_{opd}}{A_o f_{yo}} \leq 0$ or N_{opd} being as defined in Fig. 6 of section 3.115 $\gamma = 0.95\left(\dfrac{d_o}{t_o}\right)^{0.2}\left(1+\dfrac{0.0042\left(\dfrac{d_o}{t_o}\right)^{1.5}}{1+e^{(0.39q/t_o-0.53)}}\right)$	1.1

$$N_{opd} = N_1\cos\theta_1 + N_2\cos\theta_2 + N_{opg}$$

– Definition of the N_{opd} load in the chord.

Type	Formula	Failure criterion	Ultimate strength	Definition of parameters	γ_m
X, T, Y joints and K, N, KT gap joints	3.4	Shear	$N_k = \dfrac{f_{yo} t_o}{\sqrt{3}} \pi d_i \dfrac{1+\sin\theta_i}{2\sin^2\theta_i}$		1.0

Table 3.6 X, T, Y, K and N joints with I or H beam section chords

Application limits:
$30° \leq \theta \leq 90°$

$\dfrac{b_i}{t_i}$ and $\dfrac{h_i}{t_i} \leq 35$ in tension and $0.9\sqrt{\dfrac{E}{f_{yi}}}$ in compression

$\dfrac{d_i}{t_i} \leq 50$ in tension and $1.2\sqrt{\dfrac{E}{f_{yi}}}$ in compression

Type	Formula	Failure criterion	Ultimate strength	Definition of parameters	γ_m
T, Y, X joints and K and N gap joints	4.1	Plastification of the chord	$N_k = \dfrac{b_m t_w f_{yo}}{\sin\theta_i}$ where t_w is the thickness of the I or H section web	RI joints: $\left\{ b_m \leq \dfrac{h_i}{\sin\theta_i} + 5(t_o + r_o) \right.$ CI joints: $\begin{cases} b_m \leq 2t_i + 10(t_o + r_o) \\ b_m \leq \dfrac{d_i}{\sin\theta_i} + 5(t_o + r_o) \end{cases}$	1.0
	4.2	Local buckling in the chord	$N_k = \dfrac{0.8 f_{yo} t_w h_w}{\sin\theta_i}\left(\dfrac{\sigma_{cr}}{f_{yo}}\right)^{0.6}$ where h_w is the depth of the I or H section web	– needs checking for X joints if: $\dfrac{h_w}{t_w} > 1.2\sqrt{\dfrac{E}{f_{yo}}}$ – needs checking for other joints when: $\dfrac{h_w}{t_w} > 1.5\sqrt{\dfrac{E}{f_{yo}}}$ $\sigma_{cr} = \alpha E\left(\dfrac{t_w}{h_w}\right)^2$ with $\alpha = 1.44$ (X) $\alpha = 2.26$ (T, Y, K, N)	1.0
	4.3	Effective width of the bracing member	$N_k = 2f_{yi}t_i(2t_w + Ct_o)$	$C = 10.5$ for Fe 360 and Fe 430 steels $C = 7.5$ for Fe 510 steel	1.0

Table 3.6 (continued)

Type	Formula	Failure criterion	Ultimate strength	Definition of parameters	γ_m
K and N gap joints	4.4	Shear	$N_k = \dfrac{f_{yo} A_Q}{\sqrt{3}\,\sin\theta_i}$	$Q=(N_i\sin\theta_i)_{max}.\quad Q_p = \dfrac{A_Q f_{yo}}{\sqrt{3}}$	1.0
	4.5	Shear	$N_k = (A_o - A_Q)f_{yo} + A_Q f_{yo}\sqrt{1-\left(\dfrac{Q}{Q_p}\right)^2}$	$A_Q = A_o - (2-\alpha)b_o t_o + (t_w + 2r_o)t_o$ $\alpha = 0$ for CHS on I	
K and N overlap joints	4.6	Effective width of bracing member	See formula 2.6 of Table 3.4	$\alpha = \sqrt{\dfrac{1}{1+1.33\left(\dfrac{g}{t_o}\right)^2}}$ for RHS on I	1.0

θ_i : included angle between chord and bracing members
g : gap between the bracing on the face of the chord, Fig. 3.29
u : overlap, $u = q/p \times 100\%$
h_o : external 'height' of a square or rectangular chord member
h_i : external 'height' of a square or rectangular brace member (measured at 90°
      to member axis)
h_w : external 'height' of an 'I' chord
t_w : 'I' chord web thickness
r_o : Root radius, I section
β : d_i/d_o or (b_i/b_o)
N_d : design strength
N_k : characteristic strength (for the type of failure indicated in the tables)

The definition of eccentricity and its sense (positive or negative) is given in Fig. 3.28. The eccentricity must remain within limits as follows:

$$-0.55 \leqslant e/d_o \text{ or } e/h_o < 0.25$$

The definition of gap and overlap is given in Fig. 3.29. Moment caused by the eccentricity e of the joint is catered for in the tabulated values giving joint strengths, but must of course be taken into account in assessing member strengths. When the joint under consideration is subjected to both axial load and bending moment, the joint may be checked using the interaction formula:

$$F/A + M/Z \leqslant N_d/A, \quad \text{where } M/Z \leqslant N_d/2A$$

The design strength of a joint is given by N_k/γ_m, where N_k is the *characteristic strength* given in the tables, and γ_m is a reduction factor with a given value between 1 and 1.25. The cases where γ_m is assigned a value of 1 are based on a theoretical model, where actual strength is greater than calculated strength. γ_m values of 1.1 and 1.25 are assigned respectively to joints with large and small deformation capacity. The philosophy behind the development of these formulae has been to provide a reasonable margin between the maximum working load and failure load, to limit deformation to 1% of the chord width, and to ensure that there are no cracks in service conditions.

The data given may be most easily applied by means of computer program techniques, as it may be laborious to check joints by hand calculation. Hand calculation examples are supplied at the end of this section, but they are principally intended to illustrate the use of the data. The information data may be used for lattice girders, and also (where the chord type and orientation are applicable) for towers. *Node points* on 3D structures such as towers may be braced in either one plane, or adjacent planes. Nodes braced in two planes may be designed as independent plane frame joints, but the chord member must be checked for combined stresses. Tower nodes that are braced in one plane have the advantage of improved welding access.

	Yield point in N/mm²	
Thickness	$\leqslant 16$	$> 16 \leqslant 40$
Fe 360	235	225
Fe 430	275	265
Fe 510	355	345

Fig. 3.25 Minimum yield point.

f_y (N/mm²)	% ultimate elongation on the basis of 5.65 $\sqrt{A}$	$\dfrac{f_y}{\sigma_{ultimate}}$
235–450	18	$\leqslant 0.8$

Fig. 3.26

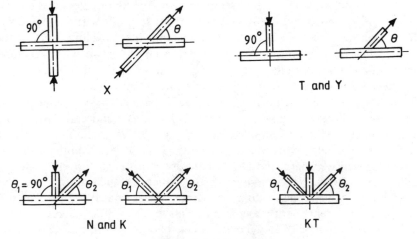

Fig. 3.27 Joint geometries.

The geometry of the joints and ease of fabrication govern the design in this case, and it is advantageous to use a β factor (brace to chord width) as close as possible to unity, especially if thin walled compression members are used for the chords. Greater joint strength can be obtained by using a higher grade of steel for the chords only, in which case the designer must ensure that the weld is correctly specified to join the two dissimilar steels. Joints with overlapped bracing are generally much stronger than those with a gap. Vierendeel type girders may only be based on the formulae given for square or rectangular chords, as information is not yet available on the behaviour of groups of 'T' joints with circle or 'I' chords.

For fabrication, fillet welds or partial penetration welds should satisfy the condition that the throat thickness should not be less than wall thickness. Alternatively, full penetration butt welds may be used. Dimensional tolerance on the setting out of adjacent braces parallel to the chord must be within 5 mm ($\pm 2\frac{1}{2}$ mm on each brace).

The authors would like to point out that great care has been taken by the research authority to ensure that all the data presented is accurate, but that the Commission for the European Communities, CIDECT, the British Steel Corporation, and the authors, assume no responsibility for errors in or misinterpretation of the information given in the diagrams or tabulated data, or in its use.

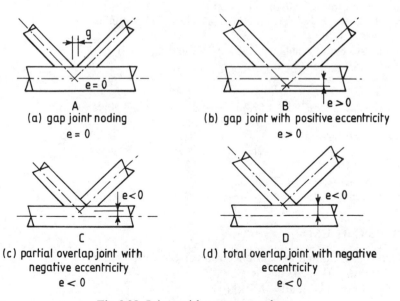

A
(a) gap joint noding
e = 0

B
(b) gap joint with positive eccentricity
e > 0

C
(c) partial overlap joint with negative eccentricity
e < 0

D
(d) total overlap joint with negative eccentricity
e < 0

Fig. 3.28 Joints with gaps or overlaps.

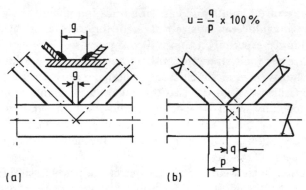

Fig. 3.29 Definition of (*a*) gap and (*b*) overlap.

3.10.2 Joint design for lattice girder with circular hollow section members

Example 3.5

Data: Ultimate load in brace $= 99\,\text{kN}$.
K joint with braces at $45°$
Steel grade 43c with $275\,\text{N/mm}^2$ yield.

Joint strength check:

(Consider the highest loaded bracing on compression chord):

Chord member 114.3 mm diameter $\times$ 6.3 mm

Brace member 60.3 mm diameter $\times$ 4 mm

$d_o/t_o = 114.3/60.3 = 18$ $g/t_o = 29/6.3 = 4.6$ $f_{yo} = 275\,\text{N/mm}^2$

Eccentricity $e = 0$

For joint strength formula use Table 3.3 (K and N joints)

$$\text{then } \gamma = 0.95(d_o/t_o)^{0.2}\left(1 + \frac{0.0042(d_o/t_o)^{1.5}}{1 + e^{(0.34g/4 - 0.53)}}\right)$$

$N_{1k} = (1.71 f_{yo}t_o^2/\sin\theta_1)(1 + 4.67\beta)\eta'\gamma$, where $\theta_1 = 45°$ and $\sin\theta = 0.707$

$N_{opd}/A_o f_{yo} \geqslant 0$ $\therefore \eta' = 1$ $t_o = 6.3$ mm (chord)

$\beta = d_i/d_o = 60.3/114.3 = 0.5275$

$N_{1k} = 1.71 \times 275 \times 6.3^2/0.707\,(1 + 4.67 \times 0.5275) \times 1 \times 2.236$

$\quad\quad = 196319\,\text{N} = 196.3\,\text{N}$

Reduction factor $\gamma_m = 1.1$ (Table 3.5)

$\therefore$ joint strength $= 196.3 \times 1.1 = 178\,\text{kN} > 99\,\text{kN}$. Hence acceptable.

3.11 PROBLEMS

3.1 Design a pinned base connection for the following data:
Column 203 $\times$ 203 UC 46
Ultimate axial load $= 170\,\text{kN}$

Ultimate shear load = 70 kN (acting on web)
Concrete grade = 20 N/mm²
(use grade 4.6 bolts)

3.2 Design a fixed base connection for the following data.
Column 305 × 305 UC 97
Ultimate axial load = 90 kN (compression)
Ultimate shear load = 30 kN (on the web)
Ultimate moment = 70 kN m (about the xx axis)
Concrete grade = 20 N/mm²

3.3 Design a fully fixed column spliced connection (using grade 8.8 bolts) for
the following data:
Column size 203 × 203 UC 46 (upper and lower column)
Ultimate axial load at splice 450 kN compression
                            180 kN tension
Ultimate $M_{xx} = 45$ kN m
Ultimate shear (xx axis) = 70 kN
No moment or shear on yy axis.

3.4 Access problems on a site limit the length of steel members to 5 m. An 8 m
long 'fixed end' beam 406 × 178 UB 74 is required. The unfactored beam
loads are 6 kN/m dead and 6 kN/m imposed. The top flange is restrained
at the centre of span. Specify the location of the splice and design the
connection using grade 8.8 bolts.

REFERENCES

1. American Institute of Steel Construction (1971) *Structural Steel Detailing* 2nd Edn.
   AISC.
2. Surtees, J. O., Gildersleeve, C. P. and Watts, C. J. (1981) A general tabular method
   for elastic and plastic analysis of eccentrically loaded fastener groups. *Structural Eng.*
   59A, June.
3. Needham, F. H. (1977) Connections in structural steelwork for buildings. *Structural
   Eng.* **58A**, September.
4. Pask, J. W. (1982) *Manual on Connections for Beam and Column Construction*. BCSA,
   London.
5. Needham, F. H. (1983) Site Connections to BS5400: Part 3. *J. Inst. Struct. Engrs.*
6. American Petroleum Institute (1982) *Recommended Practice for Planning, Designing
   and Construction of Fixed Offshore Platforms*, 13th Edn. API RP2A, American
   Petroleum Institute.
7. *Holding Down Systems for Steel Stanchions* (1980). BCSA/Concrete Society/
   Constrado.
8. *Fatigue Behaviour of Welded Hollow Section Joints* (1982) Constrado, Croydon.

9. *Design Recommendations for Hollow Section Joints* (*Predominantly Statically Loaded*) (1982). International Institute of Welding.
10. International Institute of Welding (1964) Calculation formula for welded connections subject to static loads. *Welding in the World* **2(4)**.
11. *Design of Welded Structures* (1975) The James. F. Lincon Arc Welding Foundation.

Beam to column connections 4

4.1 INTRODUCTION

Connections between beams and columns fall into two broad categories, those intended to carry moment, and those intended to carry vertical shear only. End plate connections, framed connections and seat brackets are generally designed to carry vertical shear only, as part of a 'simply supported' design. With web end plate and web framed beam connections in particular, a theory has been expounded whereby yielding of the end plate or angles should be allowed in order to reduce the moment that is transmitted from the beam to the column. This theory, based on Australian recommendations, relies on the allowance of a distance from the underside of the beam flange to the bottom of the angles or plate of not more than 33 times the plate/angle thickness, so that in theory the connection rotates about the lowest point on the plate. The authors suggest that this type of approach should only be necessary when column stresses are high (for instance, 90% of allowable) and that the successful use of end plate and framed connections in simple construction both in the UK and elsewhere has confirmed this. The thickness of end plates/angles should generally be 8 mm for sections up to and including 457 mm depth, and 10 mm for sections of greater depth. Erection shims are sometimes required when end plate construction is used, because of the lack of adjustment on the connection.

4.2 FRAMED BEAM CONNECTION (Fig. 4.1)

This type of connection is convenient when bolts only are used (no weld). Use grade 4.6 or 8.8 bolts for this type of connection. If this connection is used at both ends of a beam an effective length L_e of $1.2(L + 2D)$ applies under normal loading (code Table 9).

Example 4.1

A 356×127 UB33 has an ultimate end reaction of 75 kN, at the face of a 203×203 UC 46. Using grade 4.6 bolts and grade 43 steel, design a suitable framed connection.

SOLUTION
See Figs 4.1 and 4.2 for arrangement.
Beam web thickness $t_b = 5.9$ mm
Column flange thickness $T_f = 11.0$ mm
p_{bs} = bearing strength of connected parts

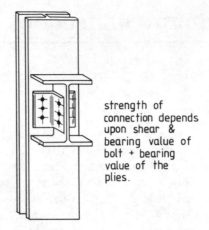

strength of
connection depends
upon shear &
bearing value of
bolt + bearing
value of the
plies.

Fig. 4.1 Framed beam connection (to resist shear).

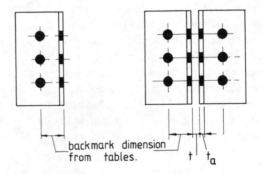

backmark dimension
from tables. t t_a

Fig. 4.2 Framed beam connection dimensions for design.

p_{bb} = bearing strength of fastener (bolt)
  d = effective diameter of fastener
  e = end distance of fastener

Consider angles to beam web connection
Minimum required value of each bolt in double shear
    $= 75/3 = 25\,\text{kN}$
Minimum required value of each bolt in bearing
    $= 75/3 = 25\,\text{kN}$
Try M16 bolts (grade 4.6) in clearance holes
Allowable value of M16 (grade 4.6) bolt in double shear
    $= 2 \times 25.1\,\text{kN}$ (from Fig. 3.1, Chapter 3)
    $= 50.2\,\text{kN} > 25\,\text{kN}$
Outer plies are 6 mm, beam web = 5.9 mm, therefore
check inner ply for bearing.

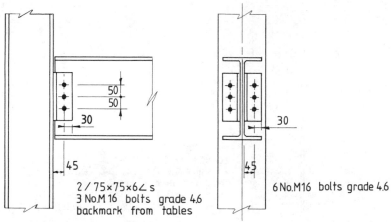

Fig. 4.3 Framed beam connection – Example 4.1 – adopted arrangement.

Bearing capacity of fastener $= dtp_{bb}$ (code 6.3.3.2 and Table 32)
$$= 16 \times 5.9 \times 435/10^3$$
$$= 41.1 \text{ kN per bolt} > 25 \text{ kN}$$

Bearing capacity of ply
$= dtp_{bs} = 16 \times 5.9 \times 460/10^3 = 43.4 \text{ kN}$ (code 6.3.3.3 and Table 33)
or $0.5 \, etp_{bs} = 0.5 \times 24 \times 5.9 \times 460/10^3 = 32.6 \text{ kN per bolt} > 25 \text{ kN}$
∴ maximum bearing capacity of M16 bolt
on the centre ply $= 32.6 \text{ kN}$

Use 3 No. M16 bolts to connect angles to beam web (Fig. 4.3)

Consider angles to column flange connection
Minimum required value of each bolt in single shear or bearing
$= 75/6 = 12.5 \text{ kN}$
Try M16 bolts in clearance holes.
Allowable value of M16 bolt in single shear $= 25.1 \text{ kN} > 12.5 \text{ kN}$
Ply capacity in bearing is adequate by inspection of bearing calculation above.

Use 6 No. M16 bolts to connect angles to column flange

and adopt arrangement as shown in Fig. 4.3 using $2/75 \times 75 \times 6$ angles.

4.3 END PLATE CONNECTION (Fig. 4.4)

This connection may extend to the full depth of the beam or only part of the depth. It has proved to be an effective and very economical method of joining beams to columns. Labour involved in fabrication consists of cutting the plate, drilling, and welding it to the beam end. Site work is limited to bolting the beam assembly to the column. If used at both ends of a beam this connection would give a beam effective length (L_e) of $1.2 \, (L + 2D)$ under normal loading (code Table 9).

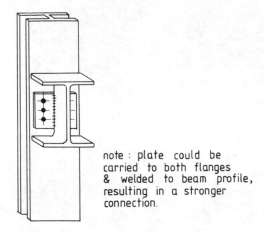

note : plate could be
carried to both flanges
& welded to beam profile,
resulting in a stronger
connection.

Fig. 4.4 End plate shear connection.

Example 4.2

Given the following information, design a suitable end plate connection using grade 43 steel, and grade 4.6 bolts.

Beam – 305 × 102UB33; Column – 203 × 203UC46; Ultimate reaction – 120 kN.

SOLUTION (Fig. 4.5)
Column flange thickness, $t_f = 11$ mm
Try a nominal end plate thickness of 10 mm (i.e. approximately equal to t_f).
Try 6 No. M16 bolts grade 4.6

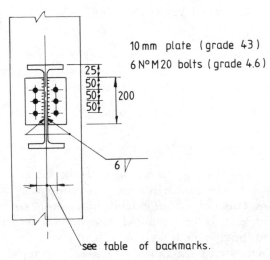

10 mm plate (grade 43)
6 N° M20 bolts (grade 4.6)

see table of backmarks.

Fig. 4.5 End plate shear connection – Example 4.2 – adopted arrangement.

Minimum single shear value per bolt $= 120/6 = 20\,kN$
Minimum bearing value per bolt $= 120/6 = 20\,kN$
Allowable value of M16 (grade 4.6) bolt
in single shear $= 25.1\,kN > 20\,kN$ (Fig. 3.1, Chapter 3)

$$\begin{aligned} \text{in bearing} &= dtp_{bb} \\ &= 16 \times 10 \times 435/10^3 \\ &= 69.6\,kN > 20\,kN \end{aligned}$$

(code 6.3.3.2)
(code Table 32)

Use 6 No. M16 bolts grade 4.6 with 10 mm end plate

Check weld on end plate
Actual vertical shear on weld $= 120 \times 10^3/(200 \times 2) - (4 \times \text{weld size})$
$$= 319\ N/mm\ \text{(assuming 6 mm weld)}$$
Allowable shear on 6 mm fillet weld (Fig. 3.8)
$$= 900\ N/mm > 319\ N/mm$$

Use 6 mm fillet weld each side of beam web

The adopted arrangement is shown in Fig. 4.5.

4.4 SEATED BEAM CONNECTION (Fig. 4.6)

This connection, which is normally intended to resist shear, should only be used when the top of the beam is supported by an angle as shown in the figure. The strength of the connection normally depends upon the ability of the seat angle to resist shear, and bending. The connection is generally used for reactions of between 50 and 140 kN (ultimate loads).

Example 4.3

A beam 533×210UB101 is connected to a 203×203UC86, using a seated beam connection. The ultimate reaction at the face of the column is 100 kN. Using grade 43 steel and grade 4.6 bolts design a suitable connection.

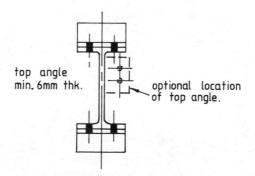

top angle
min. 6mm thk.

optional location
of top angle.

Fig. 4.6 Seated beam connection.

SOLUTION
See sketch of dimensions for design (Fig. 4.7).
K_1 = distance from centreline of root fillet to outside of beam flange.
$K_1 = (D - d)/2 = (536.7–472.7)/2 = 32$ mm
where D = overall beam depth and d = depth between root fillets
Let $a_1 = 10$ mm (nominal dimension) end clearance

Check for buckling resistance of beam web
Find p_c (allowable web stress in compression) and hence
determine P_w (max. allowable web buckling capacity).
Slenderness ratio of beam web $\lambda = 2.5\,d/t = 2.5 \times 472.7/10.9 = 109$ (code 4.5.2.1)
(where t = web thickness)
$\therefore p_c = 111$ N/mm^2 (code Table 27(c))
Try Angle $200 \times 200 \times 16$
$n_1 = K_2 + D/2 - a_1$ (Fig. 4.7) $= (18 + 16) + 536.7/2 - 10 = 292.3$ mm
where K_2 = distance from back of seating angle to the centre of its root radius
and $P_w = n_1 t_b p_c = 292.3 \times 10.9 \times 111/10^3 = 353.7$ kN > 100 kN   (code 4.5.2.1)

Check for bearing resistance of beam web
$P_{bearing} = (b + n_2)t_b p_y$
where $n_2 = 2.5 K_1 = 2.5 \times 32 = 80$ and $b = t_a + r_a - a_1 = 16 + 18 - 10 = 24$ mm
$P_{bearing} = (80 + 24) \times 10.9 \times 275 \times 10^{-3} = 311.7$ kN > 100 kN   (code 4.5.3)
   bearing adequate.

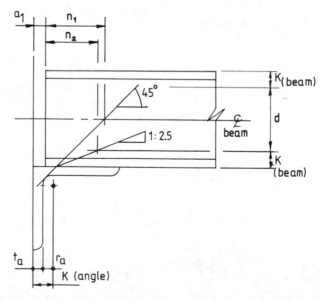

Fig. 4.7 Seated beam connection; details for checking buckling and bearing capacities of beam web.

Check bending and shear at angle root $(200 \times 200 \times 16L)$(Fig. 4.7)
e = eccentricity of load on angle
$$e = n_2/2 + a_1 - K_2$$
$$= 80/2 + 10 - 34$$
$$= 16\,\text{mm}$$
M on angle $= Re$
$$= 100 \times 0.016$$
$$= 1.6\,\text{kNm}$$
Trial minimum angle thickness $= t_a$
$$t_a = \sqrt{\frac{6M}{bxp_y}} = \sqrt{\frac{6 \times 1.6 \times 10^6}{190 \times 275}} = 13.6\,\text{mm}$$
where b = length of angle (use 16 mm leg thickness)
Use $200 \times 200 \times 16$ angle.

Average shear on angle leg $= \dfrac{\text{Shear force}}{0.9 \times \text{area leg}}$ (code 4.2.3(c))
$$= 100 \times 10^3/(0.9 \times 190 \times 16)$$
$$= 36.5\,\text{N/mm}^2 < 0.6 \times 275$$

Check weld
try 6 mm weld (E43 electrode)
Strength = 0.9 kN/mm (Fig. 3.8, Chapter 3)
Min. length of vertical weld = shear/weld strength $= 100 \times 10^3/(0.90 \times 10^3) = 111$ mm

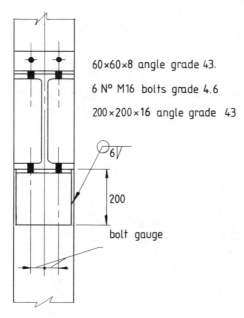

60×60×8 angle grade 43.

6 N° M16 bolts grade 4.6

200×200×16 angle grade 43

6

200

bolt gauge

Fig. 4.8 Seated beam connection – Example 4.3 – adopted arrangement.

Min. length of horizontal weld = (moment/(lever arm × weld strength)) × 2
$$= (2 \times 10^6/(200 \times 900)) \times 2$$
$$= 22.2 \, mm$$

Vertical weld provided $= 2 \times 200 \, mm - 4 \times 6 \, mm$ weld size $= 376 \, mm > 110 \, mm$. Horizontal weld provided $= 2 \times 190 \, mm - 24 = 356 \, mm > 22.2 \, mm$

Use 6 mm fillet weld all round the cleat

Final design arrangement shown in Fig. 4.8.

4.5 STIFFENED SEAT BRACKET (Fig. 4.9)

This type of connection is generally reserved for locations of high local loading, e.g. at crane gantry beams. It is usually more economical to use a beam stub instead of the seating plate and stiffener shown in Fig. 4.9. When a vertical reaction is particularly high, an additional stiffener may be found necessary.

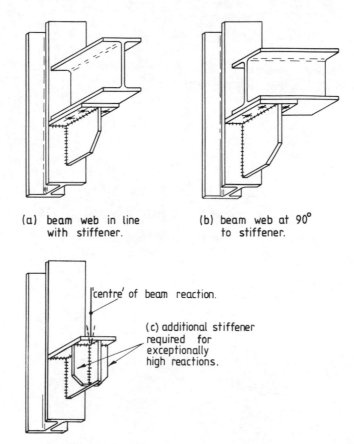

(a) beam web in line
    with stiffener.

(b) beam web at 90°
    to stiffener.

'centre' of beam reaction.

(c) additional stiffener
    required for
    exceptionally
    high reactions.

Fig. 4.9 Stiffened seat bracket – typical arrangement.

Example 4.4

Design a beam to column connection, using a stiffened seat bracket for a 457×191UB 89 with an ultimate end reaction of 150 kN carried by a 203×203UC. The beam web is at 90° (Fig. 4.9(b)) to the stiffener, and the centreline of the beam is 220 mm from the column centreline. Use grade 43 steel and E43 electrodes.

SOLUTION

With reference to Fig. 4.10:

Try preliminary sizes as follows:

Let $L_v = 250$ mm, $L_h = 215$ mm

   $t_w = 10$ mm, $t_f = 15$ mm

  $B_{col} = 203$ mm, $e = 120$ mm

    $P = 150$ kN, $a_1 = 10$ mm

Consider stiffener to act as a column,

$\lambda = d/t_w$; $d = e/\cos\theta$; $\theta = \tan^{-1}(L_v/L_h) = 49.3°$

$\lambda = 120/(0.652 \times 10) = 18.4$

Strut selection table is Table 27(b) for plate under 40 thick

whence $p_c = 273$ N/mm (allowable stress in compression)   (code Table 25)

Consider loading on stiffener

Component of axial load on stiffener $= P\sin\theta$

$$= 150/0.76 = 197\,\text{kN}$$

Eccentricity of load on stiffener $= (e - Lh/2)\sin\theta$

$$= (120 - 107.5)0.76 = 9.5\,\text{mm}$$

Area of stiffener in column action $= bt_w$

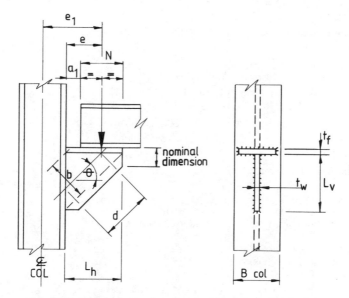

Fig. 4.10 Stiffened seat bracket – dimensions for design.

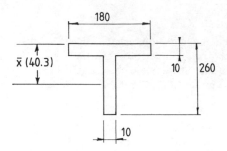

Fig. 4.11 Weld line dimensions.

where $b = L_h \sin \theta = 215 \times 0.76 = 163.4$ mm (Fig. 4.11)
Area of stiffener at 90° to the load $= 163.4 \times 10 = 1634$ mm^2
Maximum allowable stress on stiffener $= P/A \pm M$/plastic modulus
$$= (197/1634 \pm 197 \times 9.5 \times 4/163.4^2 \times 10)\,10^3 = 120 \pm 28$$
$$= 148 \text{ N/mm}^2 < 273$$
Subject to check on weld size, use 10 mm stiffener

Check weld size
Consider weld to be of unit size for stress calculation (see Fig. 4.11)
Location of neutral axis, dimension $\bar{x}$
Total weld length $= 180 \times 2 + 260 \times 2 = 880$ mm
$\bar{x} = (170 \times 10 + 260^2 \times 0.5)/880 = 40.3$ mm from top of weld.
I Weld $= bd^3/12 + Ar^2$ (neglect first term for horizontal lines of weld)
$$= 18 \times 4.03^2 + 17 \times 3.03^2 + 1 \times 21.97^2$$
$$+ 2(1 \times 26^3/12 + 26 \times 8.97^2)$$
$$= 8044 \text{ cm}^4$$
$Z_{\min}$ (bottom of weld lines) $= 8044/(26 - 4.03) = 366.1$ cm^3
$f_b \max = M/Z = Pe/Z$
$$= 114 \times 9.5 \times 10^3/366.1 \times 10^3 = 3.0 \text{ N/mm run of weld.}$$
Shear stress $= P$/No. of vert. weld lines $\times$ length
(neglect outer 10 mm vert. weld lines for shear purposes)
$$= 10^3 \times 150/2 \times 250 = 300 \text{ N/mm}$$
Resultant shear on weld $= \sqrt{300^2 + 3.0^2} = 300$ N/mm weld
$$(0.3 \text{ kN/mm})$$
Design strength of 6 mm weld $= 0.9$ kN/mm > 0.3 (Fig. 3.8, Chapter 3)
Let maximum weld size $= t_w/\sqrt{2} = 10/\sqrt{2} = 7.07 > 6$ (6 mm weld is adequate
from practical viewpoint)
Use 2 No. vertical lines of 6 mm weld to stiffener and 10 mm

stiffener (Nominal 6 mm weld to remainder of connection)

Select nominal flange thickness t_f
Note beam flange $T = 17.7$ mm
and web $t_w = 12$ mm
$\therefore$ let $t_f = 10$ mm (i.e. nominal plate thickness).
See sketch summary of design (Fig. 4.12)

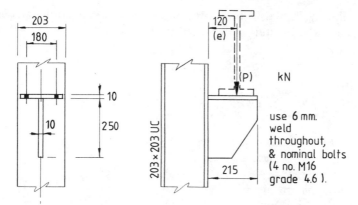

Fig. 4.12 Stiffened seat bracket – Example 4.4 – adopted arrangement.

4.6 ECCENTRICALLY LOADED CONNECTION
(To resist moment and shear)

In locations where loading cannot be placed near to the column axis, the connection must be designed for moment and shear. The two basic types of eccentrically loaded connections are:

(1) Moment in the *plane* of the connection.
(2) Moment at 90° to the plane of the connection.

The method of connection may entail bolting or welding as illustrated in Fig. 4.13, cases (a), (b), (c), (d). The connection designated (a) in this group merits particular attention. The other three cases are in general somewhat simpler to design and are illustrated by means of examples.

4.6.1 Eccentrically loaded connection case (a)
(Moment in plane of connection, bolted)

A general tabular method for elastic and plastic analysis of eccentrically loaded fastener groups was presented by Surtees, Gildersleeve and Watts, in the June 1981 edition of the *Journal of the Institution of Structural Engineers* (Tables 4.1 to 4.9). This appears to represent the best approach yet to the design of in-plane bolt groups. The method is independent of fastener type. The plastic analysis represents a fundamental condition, analogous to the plastic modulus of a beam (see Fig. 4.14(a)).

The general design objective is to use fasteners sparingly, whilst maintaining reasonable joint proportions. A choice of configurations is possible (as shown in the tables), ranging from uniform blocks of bolts to separate lines and even 'hollow' groups. The design of a bracket type joint can commence from a knowledge of the values of P, e, n_c and a. A trial value of b may be assumed and

ECCENTRICALLY LOADED BOLT GROUPS. Tables for plastic analysis (after Surtees, Gildersleeve and Watts.)

The following Tables 4.1 to 4.9 are reproduced by kind permission of the Council of the Institution of Structural Engineers:

Table 4.1

	Plastic P/nF						
	Uniform groups $n_r > 2; n_c > 2; 0.125 \leqslant p_c/p_r \leqslant 8$						
	a/b						
$e/\sum\dfrac{r}{n}$	8.0·	4.0	2.0	1.0	.50	.25	.125
5.0	.19	.19	.20	.20	.20	.20	.20
4.0	.23	.23	.24	.24	.24	.24	.24
3.0	.30	.30	.31	.31	.31	.32	.32
2.6	.33	.34	.35	.35	.36	.36	.36
2.2	.38	.39	.40	.40	.41	.42	.42
2.0	.41	.41	.42	.43	.44	.45	.46
1.8	.44	.45	.46	.46	.48	.49	.50
1.6	.48	.48	.49	.50	.52	.53	.54
1.4	.52	.52	.53	.55	.57	.59	.60
1.2	.56	.57	.58	.60	.63	.65	.66
1.0	.61	.62	.63	.66	.70	.72	.73
0.9	.64	.65	.66	.69	.74	.76	.77
0.8	.67	.68	.70	.73	.77	.79	.80
0.7	.70	.71	.73	.77	.81	.83	.84
0.6	.74	.75	.77	.81	.85	.87	.88
0.5	.77	.79	.81	.85	.89	.90	.91
0.4	.81	.83	.85	.89	.92	.94	.94
0.3	.86	.87	.90	.93	.96	.96	.96
0.2	.90	.92	.94	.97	.98	.98	.98

Table 4.2

	Plastic P/nF					
	Two rows $n_r = 2; n_c > 2; 0.125 \leqslant p_c/p_r \leqslant 8$					
	a/b					
$e/\sum\dfrac{r}{n}$	8.0	4.0	2.0	1.0	.50	.25
5.0	.20	.19	.19	.19	.20	.20
4.0	.24	.24	.24	.24	.24	.24
3.0	.31	.31	.31	.31	.31	.32
2.6	.35	.35	.35	.35	.36	.36
2.2	.40	.39	.40	.40	.41	.41
2.0	.42	.42	.43	.44	.44	.45
1.8	.45	.45	.46	.47	.48	.48
1.6	.48	.49	.50	.51	.52	.53
1.4	.52	.53	.54	.56	.57	.58
1.2	.57	.57	.59	.61	.63	.64
1.0	.62	.63	.65	.68	.70	.70
0.9	.64	.66	.68	.71	.73	.74
0.8	.67	.69	.71	.75	.77	.78
0.7	.71	.72	.75	.79	.81	.82
0.6	.74	.76	.79	.83	.85	.85
0.5	.78	.80	.83	.87	.88	.89
0.4	.82	.84	.87	.90	.92	.93
0.3	.86	.88	.91	.94	.95	.96
0.2	.91	.93	.95	.97	.98	.98

Table 4.3

	Plastic P/nF					
	Two columns $n_r > 2; n_c = 2; 0.125 \leqslant p_c/p_r \leqslant 8$					
	a/b					
$e/\sum\dfrac{r}{n}$	4.0	2.0	1.0	.50	.25	.125
5.0	.18	.18	.19	.19	.19	.20
4.0	.21	.22	.23	.23	.24	.24
3.0	.27	.28	.29	.30	.31	.32
2.6	.30	.31	.33	.34	.35	.36
2.2	.33	.35	.37	.39	.41	.42
2.0	.36	.37	.40	.42	.44	.46
1.8	.38	.40	.43	.45	.48	.50
1.6	.41	.43	.46	.50	.53	.55
1.4	.45	.47	.50	.54	.58	.60
1.2	.49	.51	.55	.60	.64	.67
1.0	.54	.57	.61	.67	.71	.73
0.9	.57	.60	.65	.71	.75	.77
0.8	.60	.63	.69	.75	.79	.81
0.7	.63	.67	.73	.79	.83	.84
0.6	.67	.71	.77	.83	.86	.88
0.5	.72	.76	.82	.87	.90	.91
0.4	.77	.81	.86	.91	.93	.94
0.3	.82	.86	.91	.95	.96	.97
0.2	.89	.92	.95	.97	.98	.98

Table 4.4

	Plastic P/nF						
	Hollow groups $n_r > 2; n_c > 2; p_c/p_r = 1$						
	a/b						
$e/\sum\dfrac{r}{n}$	8.0	4.0	2.0	1.0	.50	.25	.125
5.0	.19	.19	.19	.19	.19	.19	.20
4.0	.24	.24	.24	.24	.24	.24	.24
3.0	.30	.30	.30	.30	.31	.31	.32
2.6	.34	.34	.34	.34	.35	.35	.36
2.2	.39	.39	.39	.39	.40	.41	.42
2.0	.42	.42	.42	.42	.43	.44	.45
1.8	.45	.45	.45	.45	.46	.48	.49
1.6	.48	.48	.48	.48	.50	.53	.54
1.4	.52	.52	.53	.53	.55	.58	.60
1.2	.57	.57	.57	.58	.61	.64	.66
1.0	.62	.62	.63	.64	.68	.71	.73
0.9	.65	.65	.66	.68	.71	.74	.76
0.8	.68	.68	.69	.71	.75	.78	.80
0.7	.71	.72	.73	.75	.79	.82	.84
0.6	.75	.75	.77	.80	.83	.86	.87
0.5	.79	.79	.81	.84	.87	.90	.91
0.4	.83	.83	.85	.89	.91	.93	.94
0.3	.87	.88	.90	.93	.95	.96	.96
0.2	.91	.93	.95	.96	.98	.98	.98

Table 4.5

| | Plastic P/nF | | | | | | |
| | One column | | | | | | |
e/b n	2	3	4	5	6	7	8
5.00	.10	.07	.07	.06	.06	.06	.06
3.00	.16	.11	.11	.10	.10	.10	.09
2.00	.24	.17	.16	.15	.15	.14	.14
1.50	.32	.22	.22	.20	.20	.19	.19
1.20	.38	.28	.27	.25	.25	.24	.23
1.00	.45	.33	.32	.30	.29	.28	.28
0.80	.53	.41	.39	.37	.36	.35	.34
0.70	.58	.47	.44	.41	.40	.39	.39
0.60	.64	.53	.50	.47	.46	.45	.44
0.55	.67	.57	.53	.51	.49	.48	.48
0.50	.71	.61	.57	.55	.53	.52	.51
0.45	.74	.65	.61	.59	.58	.57	.56
0.40	.78	.70	.66	.64	.62	.61	.61
0.36	.81	.74	.70	.68	.67	.66	.65
0.32	.84	.78	.75	.73	.71	.70	.70
0.28	.87	.82	.79	.77	.76	.75	.74
0.24	.90	.86	.83	.82	.81	.80	.80
0.20	.93	.90	.88	.87	.86	.85	.85
0.16	.95	.93	.92	.91	.90	.90	.90
0.12	.97	.96	.95	.95	.94	.94	.94
0.08	.99	.98	.98	.98	.97	.97	.97
0.04	1.00	.99	.99	.99	.99	.99	.99

Table 4.6

| | Plastic P/nF | | | | | | |
| | One row | | | | | | |
e/a n	2	3	4	5	6	7	8
5.00	.09	.07	.07	.06	.06	.06	.06
3.00	.14	.11	.11	.10	.10	.10	.09
2.00	.20	.17	.15	.15	.14	.14	.14
1.50	.25	.22	.20	.20	.19	.19	.18
1.20	.29	.28	.24	.24	.23	.23	.22
1.00	.33	.33	.29	.28	.27	.27	.26
0.80	.38	.39	.35	.33	.33	.32	.32
0.70	.42	.42	.39	.37	.37	.36	.35
0.60	.45	.46	.43	.41	.41	.40	.40
0.55	.48	.48	.47	.44	.43	.43	.42
0.50	.50	.50	.50	.47	.46	.46	.45
0.45	.53	.53	.53	.50	.49	.49	.48
0.40	.56	.56	.56	.54	.52	.52	.51
0.36	.58	.58	.58	.57	.56	.55	.55
0.32	.61	.61	.61	.61	.59	.58	.58
0.28	.64	.64	.64	.64	.63	.62	.62
0.24	.68	.68	.68	.68	.68	.66	.66
0.20	.71	.71	.71	.71	.71	.71	.70
0.16	.76	.76	.76	.76	.76	.76	.75
0.12	.81	.81	.81	.81	.81	.81	.81
0.08	.86	.86	.86	.86	.86	.86	.86
0.04	.93	.93	.93	.93	.93	.93	.93

Table 4.7

$$\sum \frac{r}{n} \Big/ b$$

a/b	n_c	n_r 2	3	4	8
0.1	2	.50	.35	.34	.29
	3	.50	.35	.34	.29
	4	.50	.35	.34	.29
	8	.50	.34	.34	.29
0.2	2	.51	.37	.35	.31
	3	.51	.36	.35	.30
	4	.51	.36	.34	.30
	8	.50	.36	.34	.30
0.3	2	.52	.40	.37	.33
	3	.51	.38	.36	.32
	4	.51	.37	.36	.31
	8	.51	.37	.35	.31
0.4	2	.54	.43	.40	.36
	3	.53	.39	.38	.34
	4	.52	.39	.37	.33
	8	.52	.38	.36	.32
0.5	2	.56	.46	.43	.40
	3	.54	.42	.40	.36
	4	.53	.41	.39	.35
	8	.53	.40	.38	.34
0.6	2	.58	.49	.46	.43
	3	.56	.44	.42	.38
	4	.55	.43	.41	.37
	8	.54	.41	.39	.35
0.7	2	.61	.52	.50	.47
	3	.57	.46	.44	.41
	4	.56	.45	.43	.39
	8	.55	.43	.41	.37
0.8	2	.64	.56	.54	.51
	3	.59	.48	.47	.43
	4	.58	.47	.45	.42
	8	.56	.45	.43	.39
0.9	2	.67	.60	.58	.55
	3	.62	.51	.50	.46
	4	.60	.50	.47	.44
	8	.58	.47	.45	.41
1.0	2	.71	.64	.62	.59
	3	.64	.54	.52	.49
	4	.62	.52	.50	.47
	8	.59	.49	.47	.43

If $a/b > 1.0$ exchange a with b and n_c with n_r

Table 4.8

$$\sum \frac{r^2}{nr_c}\bigg/ b$$

a/b	n_c	n_r	2	3	4	8
0.1	2		.50	.34	.28	.22
	3		.50	.33	.28	.22
	4		.50	.33	.28	.22
	8		.50	.33	.28	.22
0.2	2		.51	.35	.29	.23
	3		.50	.34	.29	.22
	4		.50	.34	.28	.22
	8		.50	.34	.28	.22
0.3	2		.52	.36	.31	.25
	3		.51	.35	.29	.23
	4		.50	.34	.29	.23
	8		.50	.34	.28	.22
0.4	2		.54	.38	.33	.27
	3		.51	.36	.31	.25
	4		.51	.35	.30	.24
	8		.50	.34	.29	.23
0.5	2		.56	.41	.36	.30
	3		.52	.37	.32	.27
	4		.51	.36	.31	.25
	8		.50	.35	.30	.24
0.6	2		.58	.44	.39	.34
	3		.53	.39	.34	.29
	4		.51	.37	.32	.27
	8		.49	.35	.30	.25
0.7	2		.61	.47	.43	.38
	3		.54	.41	.36	.31
	4		.52	.38	.34	.29
	8		.50	.36	.31	.26
0.8	2		.64	.51	.47	.42
	3		.56	.43	.38	.33
	4		.53	.40	.36	.31
	8		.50	.37	.32	.27
0.9	2		.67	.55	.51	.46
	3		.57	.45	.41	.36
	4		.54	.42	.37	.33
	8		.50	.38	.34	.29
1.0	2		.71	.59	.55	.51
	3		.59	.47	.43	.39
	4		.55	.43	.39	.35
	8		.51	.39	.35	.30

If $a/b > 1.0$ exchange a with b and n_c with n_r

Table 4.9

Elastic P/nF

Symmetrical rectangular groups $n_r \geqslant 2; n_c \geqslant 2$

$e/\sum \dfrac{r^2}{nr_c}$	a/b						
	8.0	4.0	2.0	1.0	.50	.25	.125
5.0	.17	.17	.17	.17	.18 ·	.19	.19
4.0	.20	.20	.20	.21	.22	.23	.24
3.0	.25	.25	.26	.26	.28	.30	.31
2.6	.28	.28	.28	.30	.31	.33	.34
2.2	.31	.31	.32	.33	.36	.38	.40
2.0	.33	.34	.34	.36	.38	.41	.43
1.8	.36	.36	.37	.38	.41	.44	.46
1.6	.39	.39	.39	.41	.45	.48	.50
1.4	.42	.42	.43	.45	.49	.52	.55
1.2	.46	.46	.47	.49	.53	.58	.60
1.0	.50	.50	.51	.54	.59	.63	.67
0.9	.53	.53	.54	.57	.62	.67	.70
0.8	.56	.56	.57	.60	.65	.70	.74
0.7	.59	.59	.60	.64	.69	.74	.78
0.6	.63	.63	.64	.67	.73	.78	.81
0.5	.67	.67	.68	.71	.77	.82	.85
0.4	.72	.72	.73	.76	.81	.86	.89
0.3	.77	.77	.78	.81	.86	.90	.93
0.2	.83	.84	.85	.87	.91	.94	.96

Notation

a	is the overall width of a rectangular fastener group
b	is the overall depth of a rectangular group
e	is the eccentricity of load from the centroid of the group
F	is the limiting fastener shear
F_i	is the shear in the ith fastener
F_0	is the maximum value of F_i $(i = 1, n)$
K	is a constant determined by group geometry
M	is the applied moment on the group
M_0	is the limiting pure moment on the group
n	is the number of fasteners
n_r	is the number of fastener rows
n_c	is the number of fastener columns
P	is the applied direct force on the group (vertical alignment)
p_c	is the distance between fastener columns
P_0	is the limiting purely direct force on the group
p_r	is the distance between fastener rows
r	is the radial fastener distance from the centroid
r_c	is the radial critical fastener distance from the centroid
s_i	is the distance from the ith fastener to the instantaneous centre
s_{imax}	is the maximum value of s_i
s_0	is the limiting radius for elastic behaviour
X_{ic}	is the horizontal coordinate of the instantaneous centre with respect to the group centroid
x_i, y_i	are the horizontal and vertical distances, respectively, from the ith fastener to the group centroid
x_c	is the horizontal distance from the critical fastener to the group centroid

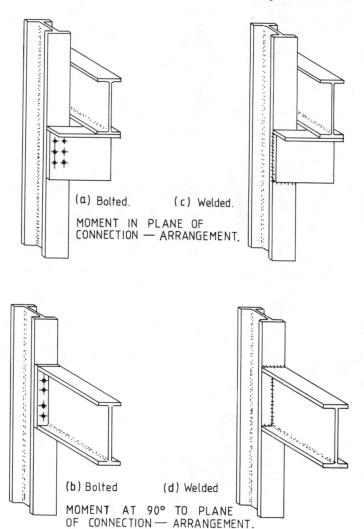

(a) Bolted.   (c) Welded.

MOMENT IN PLANE OF
CONNECTION — ARRANGEMENT.

(b) Bolted   (d) Welded

MOMENT AT 90° TO PLANE
OF CONNECTION— ARRANGEMENT.

Fig. 4.13 Eccentrically loaded connections.

$\Sigma r/n$ could then be approximately deduced from the tables. When $e/(\Sigma r/n)$ is ascertained the value of P/nF can be found. The number, size and spacing of fasteners is given by nF.

Example 4.5

Design the connection indicated in Fig. 4.14. Eccentric bracket, Case A (moment in plane of connection). Design load $P = 260\,\text{kN}$, $e = 250\,\text{mm}$. Determine the most compact fastener configuration on the basis of plastic behaviour.

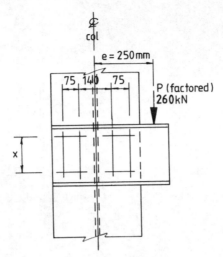

Fig. 4.14 Preliminary sketch of connection (Example 4.5.)

SOLUTION (The notation used is that given for Tables 4.1 to 4.9.)

For approx. number of fasteners, assume $a/b = 1$

Form bolt group as two column lines ($n_c = 2$) 215 mm apart, with min. vertical bolt spacing = 75 mm.

From Table 4.7 (assuming 3 rows, $n_r = 3$)

$(\Sigma r/n)/b = 0.64 = (\Sigma r/n)/a$ (a/b assumed = 1)

$\therefore e/(\Sigma r/n) = 250/(0.64 \times 215) = 1.82$

From Table 4.3:

plastic $p/nF \simeq 0.43$

$\therefore n = 260/0.43 \times F$, where F = single shear strength of bolt.

   = 11.0 bolts. Try M24 bolts at 56.5 kN per bolt.

Try three rows at 75 mm vertical spacing and check suitability as follows, with reference to Table 4.7.

$a/b = 215/150 = 1.43 > 1$ hence interchange a and b:

$b/a = 0.70$

$n_o = 2$ and $n_r = 3$ but because b and a are interchanged, n_o and n_r are interchanged.

$\therefore$ adjusted $n_c = 3$, adjusted $n_r = 2$

Thus $(\Sigma r/n)/b = 0.57$ (Table 4.7)

$e/\Sigma (r/n) = 250/(0.57 \times 215) = 2.04$

Actual $n_c = 2$ $a/b = 1.43$

Actual $n_r = 3$

and from Table 4.3, $p/nF = 0.38$, and shear value of M24 bolt = 56.5 kN

(Fig. 3.1, Chapter 3).

i.e. $p = 0.38 \times 12 \times 56.5 = 260$ kN $= P$ kN.

Use M24 bolts grade 4.6 (check on bearing capacity as shown earlier in this chapter).

The final configuration is shown in Fig. 4.16.

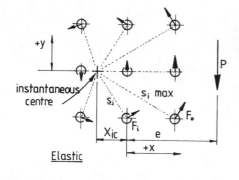

$$P/nF_o = \frac{\sum s_i^2/n\, s_i max}{e - X_{ic}}$$

where $s_i^2 = (x_i - X_{ic})^2 + y_i^2$

$$X_{ic} = \frac{\sum x_i^2 + y_i^2}{n\,e}$$

F_o = maximum fastener shear

also $F_i = \dfrac{s_i F_o}{s_i max}$

Elastic

$$P/nF = \frac{\sum s_i/n}{e - X_{ic}} \quad , \text{ all } F_i = F$$

where X_{ic} satisfies

$$e - X = \frac{\sum s_i}{\sum (x_i - X_{ic})/s_i}$$

There is no general explicit solution for X_{ic}.

Plastic

limiting radius s_o, for full force F.

As for plastic mode above, except that

$F = s_i F/s_o$ for $s_i < s_o$.

Elastic-Plastic

Fig. 4.15 Rigid plate models for eccentrically loaded fastener groups (after Surtees, Gildersleeve and Watts). Reproduced by kind permission of the Council of the Institution of Structural Engineers.

Example 4.6

Design case indicated in Fig. 4.17 (web splice).
Determine the minimum fastener number on the basis of plastic behaviour, using grade 8.8 bolts.

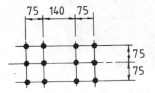

Fig. 4.16 Final bolt configuration (Example 4.5).

SOLUTION

Max. depth of bolt group = 250 mm.

Minimum number of fasteners to resist P of 480 kN using M20 grade 8.8 bolts and ignoring eccentricity = 480/91.9 = 6 (see Fig. 3.1, Chapter 3).

Max. number of fasteners in a column = 250/75 (spacing)

$$= 4$$

Try n_c (number of fastener columns) = 2, i.e. 2 each side of splice.

$a/b = 75/(3 \times 75) = 0.33$

$e = 0.5 \times 75 + 40 + 2 = 79.5$ mm.

Assuming 8 fasteners

Table 4.7 gives $(\Sigma r/n)/b = 0.37$

$$\Sigma(r/n) = 0.37 \times 225 = 83.25 \text{ (i.e. multiply by } b)$$
$$e/\Sigma(r/n) = 79.50/83.25 = 0.95$$

and from Table 4.3, $P/nF = 0.70$

hence $n = 480/(0.7 \times 91.9) = 8$

i.e. $n_r = 4$

Use columns spaced as shown on the preliminary sketch with 4 rows spaced at

75 mm, central to beam giving a total of 16 No. M20 bolts, grade 8.8.

(bearing check necessary; check in the manner shown earlier in this chapter).

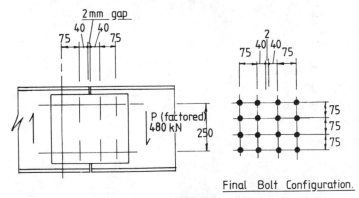

Preliminary sketch of Connection.

Final Bolt Configuration.

Fig. 4.17 Beam splice (Example 4.6).

Example 4.7

Eccentrically loaded connection, case (b) (Fig. 4.13) (moment at 90° to the plane of the connection, bolted).
Design suitable bolts for the arrangement in Fig. 4.18. The ultimate reaction R of 50 kN acts at an eccentricity of 300 mm from the face of the plate. Using a 203 × 203 UC 46 column, and a 457 × 152 UB 52 beam (grade 43 steel), design suitable bolt sizes and check plate thickness.

SOLUTION (Figs. 4.13(b) and 4.18)
l_a = bolt group moment arm = distance between centrelines of bolt sub-groups

$$= 2\left(75 + \frac{75}{2}\right)$$

$$= 225 \text{ mm}.$$

Vertical shear per bolt $= 50/8 = 6.25$ kN
M = Moment on bolt group $= Rl_a = 50 \times 0.225$ kNm
$\quad$ Assumed tension per bolt $= M/l_a n = 50 \times 0.225/(0.225 \times 4) = 12.5$ kN
where n = number of bolts in sub-group
Tension capacity of one M16 bolt, grade 4.6 = 30.6 kN (Fig. 3.1, Chapter 3)
Use M16 grade 4.6 bolt arranged as shown in Fig. 4.18

Design end plate
Check bearing capacity of fastener and connected ply.

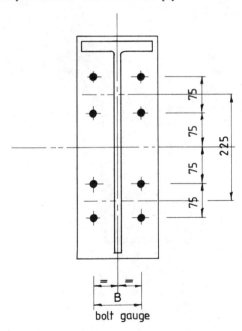

Fig. 4.18 Moment at 90° to the plane of connection (bolted).

Bearing capacity of fastener $= dtp_{bb}$ (code 6.3.3.2 and Table 32)
$$= 16 \times 10 \times 435/10^3$$
$$= 69.6\,\text{kN} > 6.25\,\text{kN}$$

where $t =$ plate thickness, to be approximately equal to column flange thickness, and taken as 10 mm.

Bearing capacity of ply $= dtp_{bs}$ or $0.5etp_{bs}$ (code 6.3.3.3)

where $p_{bs} = 460$ N/mm^2 and $e =$ end distance of fastener   (code Table 33)

$\therefore$ bearing capacity $= 0.5 \times 24 \times 10 \times 460/10^3 = 55.2\,\text{kN} > 6.25$

Check plate bending capacity
(between pull of beam web & line of bolts)
Tension per bolt $= 12.5$ kN
Assume load to spread at approximately 45°. Therefore
width of loaded web $= 75$ mm.
(For larger connections see Section 4.6.3.)
$M =$ tension $\times B/4 = 16.67 \times 0.09/4 = 0.375$ kNm
where $B =$ bolt gauge $= 90$ mm

Min. S reqd $= \dfrac{M}{f} = \dfrac{0.375 \times 10^6}{275 \times 10^3} = 1.36\,\text{cm}^3$

Min. plate thickness $t = \sqrt{1.36 \times 6/7.5} = 1.04$ cm

Use 12 mm thick plate

For design of weld see Example 4.9.

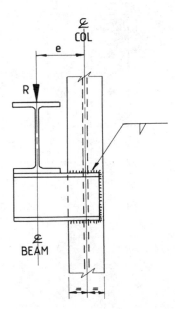

Fig. 4.19 Moment in plane of connection (welded).

Example 4.8

Eccentrically loaded connection, case (c) (Fig. 4.19) (moment in plane of connection using weld).
Design a connection of this type given the following data:
$R = 10$ kN ultimate load (E43 welds and grade 43 steel)
$e = 200$ mm eccentricity
$B = 100$ mm length of flange welds $\Big\}$ (see Fig. 4.18)
$D = 450$ mm length of web weld

SOLUTION (Fig. 4.20)
In the absence of available research data on the plastic design of welds eccentrically loaded in the plane of the connection, the authors suggest that for the present, welds should be designed by the 'elastic' method as follows:

Weld centre of gravity
$\bar{y} = \Sigma ay/A$ where $A = 450 + (2 \times 100) = 650$ and $a =$ flange weld length $= 100$ mm
Hence $\bar{y} = 2(100 \times 50)/650 = 15.38$ mm
$I \,(\text{Polar}) = I_x + I_y$
$$I_x \,(\text{weld}) = \frac{bd^3}{12} + Ar^2$$
$$= (1 \times 450^3/12) + 2(100 \times 225^2 + 100 \times 1^3/12)$$
$$= 17.7 \times 10^6 \text{ units}^4$$
$$I_y \,(\text{weld}) = \frac{bd^3}{12} + Ay^2$$
$$= (450 \times 1^3/12 + 450 \times 15.38^2)$$
$$+ 2(1 \times 200^3/12 + 200 \times 84.62^2)$$
$$= 4.3 \times 10^6 \text{ units}^4$$
$$I \,(\text{Polar}) = (4.3 + 17.7) \, 10^6 \text{ units}^4$$
$$= 22 \times 10^6 \text{ units}^4$$

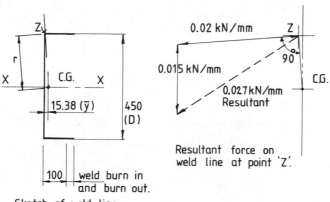

Fig. 4.20 Moment in plane of connection (welded). Solution by drawing.

Vertical load per mm of weld $= 10\,\text{kN}/650\,\text{mm}$
$$= 0.015\,\text{kN/mm}$$

Moment on weld line $= Pe$
$$= 10 \times 200$$
$$= 2000\,\text{kN/mm}$$

Polar force at point 'Z' $= M\bar{x}/I$ (polar)
$$= 2000 \times 225/22 \times 10^6$$
$$= 0.02\,\text{kN/mm}$$

Resultant maximum force on
weld at point 'Z' $= 0.027\,\text{kN/mm} < 0.7\,\text{kN/mm}$ for 5 mm fillet weld

The same result can be obtained by drawing as shown in Fig. 4.20.

Use 5 mm fillet weld throughout

Example 4.9

Eccentrically loaded connection Case D (Fig. 4.21) (moment at 90° to the plane of connection using weld).

Design the connection using grade 43 steel and E43 electrodes given the following data:

$R = 50\,\text{kN}$ ultimate load

$e = 300\,\text{mm}$

Beam – $305 \times 102\text{UB}28$

Column – $152 \times 152\text{UC}37$

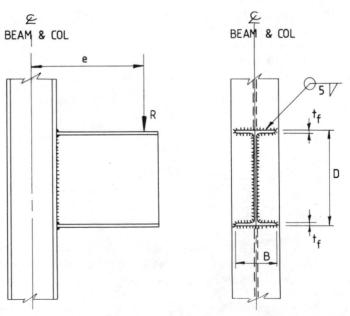

Fig. 4.21 Moment at 90° to plane of connection (welded). Dimensions for design.

SOLUTION

See Fig. 4.21 for dimensions for design.

S = Plastic modulus of weld line; assuming line to be of unit width

$$S = BD + (B - t_w)d + \frac{d^2}{4}$$

where $B = 101.9$ mm, $D = 308.9$ mm, $t_w = 6.1$ mm and $d = D - 2t_f = 308.9 - 17.8 = 291.1$ mm

$$\therefore S = (30.89 \times 10.19) + (10.19 - 0.61)29.11 + 29.11^2/4$$
$$= 805.5 \text{ cm units}$$

Value of M

$M = Re = 50 \times 0.3 = 15$ kNm

Stress f_b (find in terms of kN per mm)

$f_b = M/S = 15 \times 10^3/805.5 \times 10^3 = 0.02$ kN/mm of weld at the extreme fibres

Strength of 5 mm fillet weld $= 0.75$ kN/mm > 0.02 (Fig. 3.1, Chapter 3)

Use 5 mm profile fillet weld for connection

4.7 BEAM TO COLUMN MOMENT CONNECTIONS

There are two general forms for this type of construction, i.e. semi-rigid or fully rigid. Semi-rigid connections are used where minor moments are likely to occur, and where the imposed moment is much less than the capacity of the beam. Rigid connections are used with higher moments such as those that occur with plastic design of portal frames. Stiffeners are likely to be required in this latter case. Some typical types of connection are illustrated in Figs 4.22(a)–(d) and 4.23(e)–(h) and can be briefly described as follows:

(a)   Semi-rigid connection   Major moment in one direction only.

(b)/(c)   Rigid connection   Major moment may be reversible.

(d)   Rigid connection   Major moments in one direction only, use of T sections may obviate lamellar tearing that could possibly occur with use of thick welded plate connections.

(e)/(f)   Rigid connection   Major moments in one direction only. Apply the usual checks for stiffeners and column web buckling.

(g)/(h)   Rigid connections   Methods of connecting beams to column webs so that a moment can safely be developed in the weak axis of the column.

Where fully rigid design is used, the connection is usually expected to develop the full beam moment capacity. When full stresses are used in these cases, problems arise of bending of the end plate or column flange plate, beam web buckling or excessive web bearing. In addition to this, the column web must be checked for bearing and buckling, and stiffeners provided if necessary. The column web should also be checked for shear, and if necessary web 'doubler' plates, or diagonal stiffeners, added. Depending on the amount of restraint

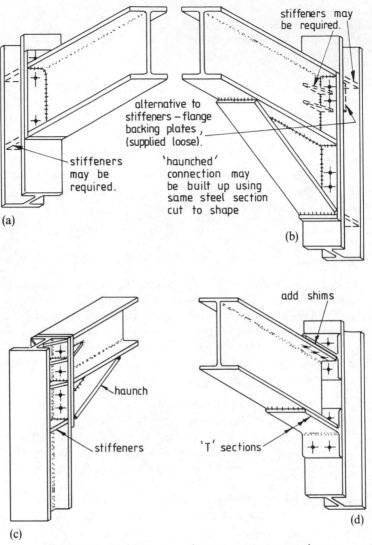

stiffeners may
be required.

alternative to
stiffeners – flange
backing plates,
(supplied loose).

stiffeners
may be
required.

'haunched'
connection may
be built up using
same steel section
cut to shape

(a)

(b)

haunch

stiffeners

'T' sections

add shims

(c)

(d)

Fig. 4.22 Typical beam to column moment connections.

offered to the beam ends, an effective length L_e can be assumed of either $0.85\,L$ or $0.7\,L$ for normal loading (code Table 9).

4.7.1 The use of bolts in moment connections

The type and arrangement of bolts in a moment connection will determine the design method to be used. A connection for ordinary bolts where the connection

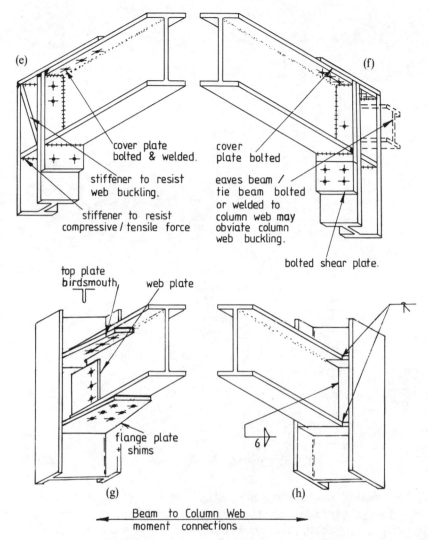

cover plate
bolted & welded.

stiffener to resist
web buckling.

stiffener to resist
compressive / tensile force

cover
plate bolted

eaves beam /
tie beam bolted
or welded to
column web may
obviate column
web buckling.

bolted shear plate.

top plate
birdsmouth,

web plate

flange plate
shims

(g)   (h)

Beam to Column Web
moment connections

Fig. 4.23 Typical beam to column moment connections.

substantially consists of relatively long bolt lines, should be designed by assuming
a neutral axis a distance n up from the underside of the beam, as indicated in
Fig. 4.24 (following practice recommended by the AISC). If, however, HSFG
bolts are used, the neutral axis should be assumed to be at the centroid of the
whole bolt group; this allows for the clamping effect of HSFG bolts.

Bolts are often located inside the depth of haunched connections for clearance
reasons or for architectural appearance. Bolts are more efficiently used by
grouping them both sides of the tension flange, a common arrangement being

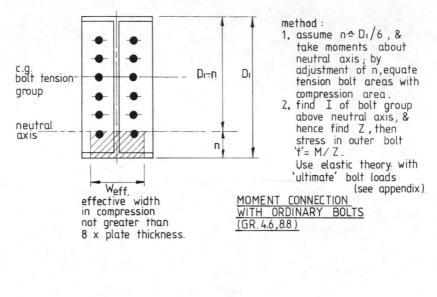

method :
1. assume $n \doteq D_1/6$, & take moments about neutral axis ; by adjustment of n, equate tension bolt areas with compression area .
2. find I of bolt group above neutral axis, & hence find Z , then stress in outer bolt 'f'= M/ Z .
Use elastic theory. with 'ultimate' bolt loads
(see appendix).

c.g.
bolt tension
group

neutral
axis

D_1-n D_1

n

W_eff.
effective width
in compression
not greater than
8 x plate thickness.

MOMENT CONNECTION
WITH ORDINARY BOLTS
(GR. 4.6, 8.8)

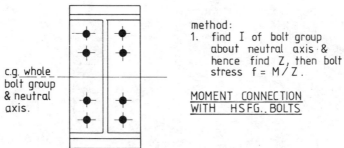

c.g. whole
bolt group
& neutral
axis.

method:
1. find I of bolt group about neutral axis & hence find Z, then bolt stress f = M / Z .

MOMENT CONNECTION
WITH HSFG.. BOLTS

Fig. 4.24 Bolted moment connections.

two bolts outside the beam depth and the remainder of the group inside. Bolts grouped near the tension flange may be designed by assuming that the connection effectively rotates about the centreline of the compression flange. In such cases, ordinary bolts may be designed by placing the neutral axis at the compression flange centreline and obtaining the elastic modulus of the bolt group. Thus, the highest stressed bolts would be those most distant from the compression flange and the force in each bolt P = moment/modulus Z of the bolt group (let each bolt have an area value of unity, then Z about the compression flange = $\Sigma Ay/y_{max}$).

For the same arrangement of fasteners, using HSFG bolts, an allowance for the clamping action of the bolts must be made. This is usually done by assuming that the bolts act at the same tension, effectively at the centroid of the group, so that the force in each HSFG bolt $P = M/l_aN$, where l_a = lever arm from the

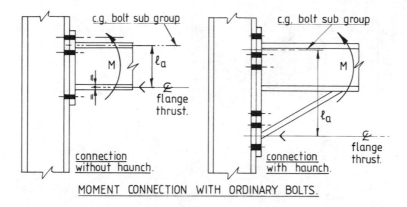

In each case, Bolt tension = Moment divided by the product of the lever arm (l_a) and the number of bolts in the sub-group under tension.

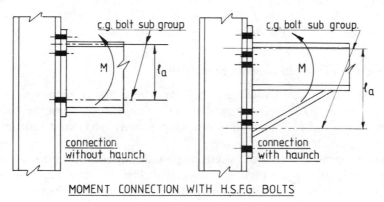

Fig. 4.25 Bolted moment connections.

centroid of the bolt group to the centreline of the compression flange and $N =$ no. of bolts (see Fig. 4.25).

The restraint offered by the remainder of the connections outlined in this chapter is such that the beam effective length L_e is $0.7 L$ under normal loading conditions (both ends of beam similarly connected).

4.7.2 The use of welds in moment connections

As a general rule one may assume that vertical shear is carried by the beam web and moment by the beam flanges, the latter being translated into a force

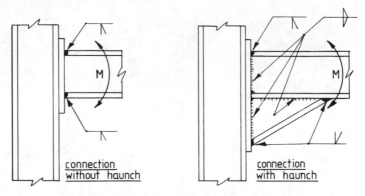

Fig. 4.26 Welded moment connections (major moments).

couple acting at a lever arm (l_a) between the flange centrelines. If the moment is minor, use a uniform size fillet weld for the beam to plate connection, but for major moments use full strength butt welds at the flange/plate interface, especially if a haunch is used (see Fig. 4.26). When moments are high and end plates are thick (greater than 20 mm) the possibility of lamellar tearing exists; for design guidance on this subject see Chapter 1.

4.7.3 Use of stiffeners in moment connections

Horizontal stiffeners are used in locations where the column web would otherwise buckle, or fail in bearing, due to the flange forces from the beam or haunch (see Fig. 4.27). A method of checking whether horizontal stiffeners are required is given later in this chapter.

Diagonal stiffeners are used where large shear forces are imposed on the column web locally due to the applied moment. The design of such stiffeners is given later in this chapter. Alternatively, it is possible to strengthen the column web locally by welding 'doubler' plates each side.

Vertical stiffeners are sometimes used on beams at the end of a haunch, if the calculated force from the haunch would otherwise cause web buckling or bearing failure. End plate and column flange stiffeners are used where the plate or flange would otherwise bend excessively. A recent alternative to column flange stiffeners is the use of flange backing plates; supplied loose, these plates prevent local overstressing of the column flange. In the absence of suitable data on flange backing plates, the authors suggest that the plates could be sized in thickness by assuming elastic action on the plate and the column flange, using the elastic Z values to check stress. The force that plate and column flange carry would be in proportion to the I values, that is (plate thickness)^3: (column flange thickness)^3.

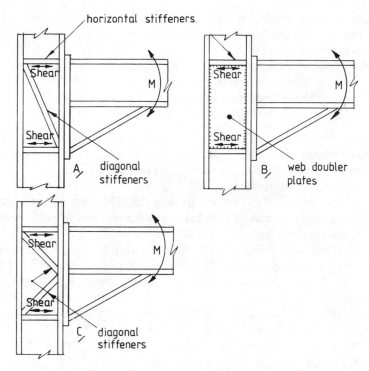

Fig. 4.27 Prevention of local column web buckling.

4.7.4 Procedure for design of a beam/column moment connection
(see also Example 4.10)

(1) Determine a range of loading cases at the connection, noting down, for each case, moment, shear, and axial load (tension or compression).
(2) Having numbered the loading cases, select the apparently worst case and, if necessary, check against the other cases when a final design has been completed for this case.
(3) Sketch a preliminary connection based on experience and the client's special requirements, if any. If without experience, refer to Example 4.10 for guidance.
(4) Decide on the bolt configuration, e.g. try, for instance, two groups of four bolts. Avoid using more than one line of bolts either side of a web for a moment connection. *If four lines of bolts are used,* employ triangular stiffeners to ensure that the outer line of bolts carry load, otherwise all the load would be carried by the inner lines and their capacity could be exceeded.
(5) Determine basic tension in the outermost bolt pair. See Fig. 4.24 for sketch of

'tension' for ordinary and HSFG bolts. Determine bolt size based on maximum values of tension and shear per bolt for this connection.
(6) Design end plate weld
   (a) Assume moment is translated into force in each flange. Design flange weld for tension or compression.
   (b) Assume shear force is carried by web. Design web weld.
   (c) Adopt uniform weld size for all fillet welds to connection.

(7) Having determined flange force as noted in 6(a) above, check column web buckling (code 4.5.2.1 and Example 4.13). Add column stiffeners if necessary.
(8) If column horizontal web stiffeners are omitted
   (a) Assume an end plate size.
   (b) Check column web bearing (code 4.5.3). If bearing is too large add horizontal stiffeners or web doubler plates on both sides of column web (same thickness).
   (c) Check column flange and end plate for local bending. If bending is too large add triangular and horizontal stiffeners as necessary; or for end plate increase thickness, and for column flange add flange backing plates.

(9) Check shear on column web. (Shear capacity of a web $p_v = 0.6p_y dt$ – code 4.5.3. See also Fig. 4.28). If web is not sufficiently thick, use diagonal stiffeners, or web doubler plates to improve V_{cr} (see code 4.4.5.3).
(10) Highlight all relevant data in your calculation for the guidance of the detailer, supplying a summary sketch at the end of the calculation.

   The following examples of beam to column connections are given to illustrate the type of consideration that normally applies and the methods that may be used to arrive at a reasonable solution.

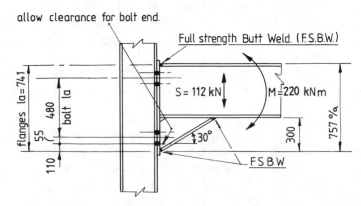

Fig. 4.28 Trial connection using HSFG bolts (Example 4.10).

4.8 BEAM/COLUMN MOMENT CONNECTIONS – DESIGN EXAMPLES

Example 4.10

Given a beam to column moment connection as described below, design the connection, welds and bolts. Note any further checks to be made.
Connection data:
Beam to column moment 220 kNm (ultimate) *reversible.*
Beam to column shear 112 kN (ultimate)
Beam: 457 × 191 UB 67
Column: 305 × 305 UC 97
Maximum haunch depth = 350 mm
Use grade 43 steel, HSFG bolts, E43 electrodes

SOLUTION
Worst case, ultimate moment and shear as noted above. See Fig. 4.28 for trial connection.
Note: (A) Haunch introduced to increase bolt
          lever arm (cut from 457 × 191 UB 67, same section as beam)
       (B) Bolts kept inside depth of beam and haunch for architectural reasons.
Tension in bolt sub-group = M/bolt l_a = 220/0.48 (say) = 458.3
(check that bolt lever arm is not less than this value in final arrangement)
Tension T per bolt = 458.3/4 = 114.5 kN
Shear per bolt = 112/8 = 14 kN
Use M20 HSFG bolts
Let slip factor = 0.45 (Connection interface to be left in unpainted, wire brushed condition – see Section 3.2, Chapter 3).

End plate thickness t
Assume force on bolt to be carried through plate as illustrated in Fig. 4.29. Check plate

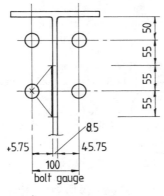

assumed plate bending

Fig. 4.29 View on top portion of beam end plate.

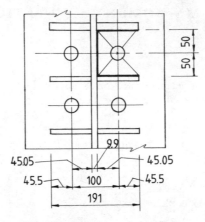

Fig. 4.30 View on column flange.

for bending on perimeter, making simplified assumption that the load is carried uniformly to the perimeter.

Length of perimeter shown $= 2 \times 55$ mm $= 110$ mm (worst case)

Tensile force in bolt $= 114.5$ kN

Maximum distance to beam flange/web $= 50$ mm

Moment in plate at beam/web $= 114.5 \times 0.05/2$ (double curvature bending)
$$= 2.86 \text{ kNm}$$

Plastic modulus $S = M/f = bt^2/4$ so that $t = \sqrt{(4M/fb)}$

Min. plate thickness (let $p_y = 265$ N/mm^2; code Table 6)

$t = (4 \times 2.86 \times 10^6/265 \times 110)^{1/2} = 19.8$ mm.

Use 20 mm plate

Column flange thickness

Column flange is 15.4 mm (less than idealized end plate thickness of 19.8 mm). Therefore add triangular stiffeners, nominal 10 mm thickness (check by assuming column action of the stiffeners described in Section 4.5, Example 4.4).

Check that column flange is adequate with stiffening (see Fig. 4.30).

Length of perimeter $= 90.55 \times 2 + 100 = 281.1$ mm

Tensile force in bolt $= 114.5$ kN

Max. distance to plate stiffener/col. web $= 50$ mm

Moment in plate at flange/web $= 114.5 \times 0.05/2 = 2.86$ kNm

Let $p_y = 275$ N/mm^2 (code Table 6).

$\therefore d = (4 \times 2.86 \times 10^6/275 \times 281)^{1/2} = 12.16$ mm (i.e. less than 15.4 mm; therefore column flange is adequate in thickness with the stiffeners shown).

End plate weld

Flange force $= M/\text{lever arm} = 220/0.741 = 297$ kN

Breadth of beam $= 189.9$ mm

force on full strength butt weld per mm $= 297/189.9 = 1.6$ kN mm

clearly a wide margin of safety.

Use FSBW to haunch ends and top beam flange

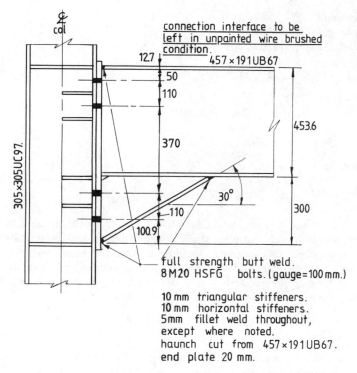

Fig. 4.31 Beam/column moment connection – Example 4.10.

Shear Force on web = shear/depth = $112/(715 \times 2) = 0.078\,\text{N/mm}$
Allowable force on 5 mm fillet weld = $0.75\,\text{kN/mm} > 0.078$

Use 5 mm fillet weld throughout (except where noted – Fig. 4.31)

Further checks to be carried out: Check column web shear, as shown in following examples, and carry out all the above checks where necessary, for the case of moment reversal.

Example 4.11

A failure mechanism is assumed for the bolted end plate in Example 4.10 whereby the plate yields in clearly defined planes at the angles θ_1 and θ_2 (see Fig. 4.31).

Given the data for the connection designed in Example 4.10, check the beam end plate thickness using the yield line theory.

SOLUTION
Internal work, W_i
$$W_i = M_y[210\theta_1 + 110\theta_1 + (91\theta_2 \times 2) + (45\theta_2 \times 2) + (50\theta_1 + 46\theta_2)2]2$$
$$= 2M_y\,(420\theta_1 + 364\theta_2) \tag{1}$$

By inspection of Fig. 4.32 $50\theta_2 = 46\theta_1$

Substituting in (1) gives $W_i = 1631\,M_y\theta_1$ (2)

The plastic moment capacity per unit length is given by

$M_y = 0.25\,t^2\sigma_y = t^2 \times 275/4 = 68.75t^2$

Substituting in (2) gives $W_i = 112131t^2\theta_1$

External work

$W_e = P_c\delta =$ Force (in bolts) $\times$ displacement

(Example 4.10)

$= 458.3 \times 10^3 \times 46\theta_1$

Energy balance

$W_i = W_e$

$112131t^2\theta_1 = 458.3 \times 10^3 \times 46\theta_1$

$\therefore t = 13.7\,\text{mm}$

Using this method a 15 mm thick end plate is shown to be adequate, resulting in significant saving in both end plate thickness and in stiffeners, compared to the size arrived at in Example 4.10. Stiffeners should be added at the lower horizontal yield line. In order to arrive at a realistic design the diagonal between θ_1 and θ_2 should be about 45°, and the distance from the vertical edge of end plate to the centreline of bolts should be less than the centreline of bolts to the edge of the web of the beam.

Example 4.12

Given a beam to column connection as described in Example 4.10, check the beam end of the haunch for web buckling and bearing and, if necessary, add a stiffener.

SOLUTION

Beam flange force (from Example 4.10) $= 297\,\text{kN}$

Angle of haunch $= 30°$

Vertical component of force from haunch $= 297 \tan 30° = 171.5\,\text{kN}$

Contact length of flange $b_1 = 12.7/0.577 = 22\,\text{mm}$

Check bearing:

$k =$ flange thickness $+$ root radius $= 22.9\,\text{mm}$

$t =$ web thickness $= 8.5\,\text{mm}$

Length in bearing $= 22 + 2.5k = 79\,\text{mm}$ (code 4.5.3)

Bearing capacity $= 79tp_y = 79 \times 8.5 \times 275/10^3 = 185\,\text{kN} > 171.5\,\text{kN}$

Check buckling of web:

Depth of web between root fillets, $d = D - 2k = 453.6 - 2 \times 22.9 = 407.8\,\text{mm}$

Buckling resistance $P_w = (b_1 + D)tp_c$ (code 4.5.2.1)

$= (22 + 453.6)8.5p_c$

Slenderness $\lambda = 2.5d/t = 2.5 \times 407.8/8.5 = 120$

From code Table 27(c), for $p_y = 275\,\text{N/mm}^2$, $p_c = 97\,\text{N/mm}^2$

$\therefore P_w = 475.6 \times 8.5 \times 97/10^3 = 392\,\text{kN} > 171.5\,\text{kN}$

Hence, vertical stiffeners are not required at the haunch end.

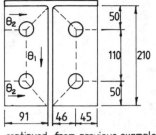

continued from previous example.

Fig. 4.32 Bolted end plate to yield line theory.

Example 4.13

Design of horizontal stiffeners at end plate/column flange connection
Given a beam to column moment connection as described below, check the connection
for web buckling and bearing and design the stiffeners.
Beam to column moment = 220 kNm (ultimate)
Beam to column shear = 112 kN (ultimate)
Beam: 457 × 191 UB 67
Column: 305 × 305 UC 97
Haunch depth = 300 mm
End plate thickness = 20 mm
Grade 43 steel, HSFG bolts, E43 electrodes.

SOLUTION
Check whether horizontal stiffeners to column web are required.
Flange force = 297 kN   (see Example 4.10)
Force in bearing on column web   (see Fig. 4.33)
$$= \frac{297 \times 10^3}{205.95 \times 9.9} = 145.7 \, \text{N/mm}^2$$ (code 4.5.2.1)
$$(< p_y = 275 \, \text{N/mm}^2)$$
Length of web in compressive buckling (45° load spread)
$$= 12.7 \, \text{mm} + D \, (\text{column}) + 2t_e$$
$$= 12.7 + 307.8 + 40$$
$$= 360.5 \, \text{mm}$$
Force on web in compressive buckling
$$= 297 \times 10^3/360.5 \times 9.9 = 83.6 \, \text{N/mm}^2$$
Slenderness of column web $\lambda = 2.5 \, d/t$ (code 4.5.2.1)
$$= 2.5 \times 246.6/9.9 = 62.2$$
From code Table 25, for flat bar under 40 mm thick, the strut table to use is Table 27(b)
whence $p_c = 216.6 \, \text{N/mm}^2 > 83.6 \, \text{N/mm}^2$
Horizontal stiffeners to the column web are therefore not required.
Check for end plate and/or column flange stiffeners.
End plate thickness = 20 mm > 15.4 mm (column flange)
Add local triangular stiffeners to the column flange, making the plate span between

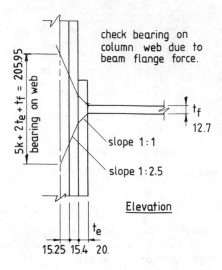

check bearing on
column web due to
beam flange force.

$5k + 2t_e + t_f = 205.95$

bearing on web

t_f

12.7

slope 1:1

slope 1:2.5

Elevation

t_e

15.25 15.4 20.

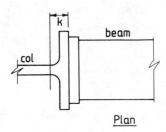

k

beam

col

Plan

Fig. 4.33 Example 4.13.

stiffeners and column web. Another option is to add flange backing plates as mentioned in Section 4.7.3 of this chapter.

Check for shear resistance of column web
Shear on column web = flange force from beam = 297 kN.
Shear buckling resistance $V_{cr} = dtq_{cr}$ (code 4.4.5.3)
$d/t = 24.9$ so from code Table 21(b), for $p_y = 275 \, \text{N/mm}^2$
$q_{cr} = 165 \, \text{N/mm}^2$
and $V_{cr} = 246.5 \times 9.9 \times 165/10^3 = 402.7 \, \text{kN} > 297 \, \text{kN}$
∴ web is adequate.
For final connection see Fig. 4.31

Note: horizontal stiffeners have been added to the top and bottom of the connection because triangular stiffeners to the column flange (see Example 4.10) would be required in this location in any case, and the arrangement chosen has a cleaner appearance.

Example 4.14

Check for horizontal and diagonal stiffeners
Given the data below, check the beam/column moment connection for column web buckling and bearing, and design the stiffeners.
Connection fully welded
Beam to column moment = 350 kNm (ultimate)
Moment in one direction only, with top flange in tension.
Beam to column shear = 80 kN.
Beam: 457 × 191 UB 67
Column: 305 × 305 UC 97
Haunch depth = 300 mm
Grade 43 steel, E43 electrodes

SOLUTION
Check for horizontal stiffeners to column web
Flange force = ultimate moment/lever arm = 350/0.441 = 794 kN
(Lever arm = distance between centres of column flanges)
Force in bearing on column web
= $794 \times 10^3/(5k + t_f) \times t_w$
= $794 \times 10^3/(5 \times 30.6 + 12.7)9.9 = 484 \,\text{N/mm}^2 > 275$.
k = column flange thickness + root radius and t_f = beam flange thickness.
The local capacity of the web is exceeded, so use horizontal stiffeners. Try stiffeners 10 mm thick. Check capacity of stiffeners and column web.
Radius of gyration of composite strut of stiffener and column web, Fig. 4.34:
I value about axis parallel to web = $bd^3/12$
$\simeq 10 \times 304.8^3/12 \times 10^4$ (neglecting I of web about longitudinal axis)
= $2360 \,\text{cm}^4$

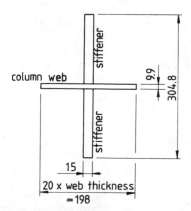

Plan on composite strut of
stiffener and column web.

Fig. 4.34 Example 4.14.

Minimum area of stiffener required $= 0.8\,F_x/p_{ys}$ (code 4.5.4.2)
$$= 0.8 \times 794 \times 10^3/275 \times 10^2$$
$$= 23.1\,\text{cm}^2$$
Actual area, $A = (304.8 \times 10 + 183 \times 9.9)/10^2 = 48.6\,\text{cm}^2 > 23.1\,\text{cm}^2$
$$r = \sqrt{(I/A)} = \sqrt{(2360/48.6)} = 6.97\,\text{cm}$$
Length of stiffener $L_E = \text{col.depth} - 2 \times \text{col.flange thickness}$
$$= 307.8 - 2 \times 15.4 = 276.6\,\text{mm}$$
Slenderness ratio of stiffener $= 276.6/69.7 = 3.96$
From code Table 25 the strut table to use is Table 27(b)
$\therefore$ allowable max.stress $= 275\,\text{N/mm}^2$
Maximum allowable load on horizontal stiffeners
$= p_y \times \text{area} = 275 \times 4860/10^3$
$= 1336.5\,\text{kN} > \text{actual load of } 794\,\text{kN}$
Use 10 mm thick horizontal stiffeners as shown in Fig. 4.34

Check allowable weld size.
Try 5 mm weld.
Allowable force on 1 mm of weld $= 5 \times 0.7 \times 1 \times 215/10^3$ (code Table 3.6)
$$= 0.75\,\text{kN/mm}$$
Minimum 5 mm weld length $= 794/0.75 \times 4$ (weld runs)
$$= 265\,\text{mm (length to be provided} = 276\,\text{mm)}$$

Use 5 mm weld each side of stiffeners

Check for diagonal stiffeners to column web
Flange force in beam $= 794\,\text{kN}$
Shear buckling resistance of column web   (from Example 4.14)
$= 402.7\,\text{kN} < 794\,\text{kN}$. This is inadequate so

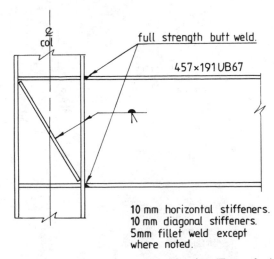

Fig. 4.35 Beam/column moment connection (Example 4.14).

try 10 mm diagonal stiffeners each side of column web.

Load to be carried by stiffeners $= 794 - 402.7 = 391.3$ kN horizontal shear

Resultant axial force on stiffeners $= 391.3$ kN/cos θ

where θ is angle of stiffeners to the horizontal

$= \tan^{-1}(453.6/307.8) = 55.84°$

$\therefore$ resultant axial force $= 266$ kN.

By inspection, 10 mm diagonal stiffeners are adequate

If desired, check by assuming the stiffeners to act as a small column with axial load, including a nominal width of column web in the calculation.

The arrangement of stiffeners determined is shown in Fig. 4.35.

4.9 PROBLEMS

4.1 A 305 × 165UB40 has an ultimate end reaction of 90 kN, at the face of a 203 × 203UC46. Using grade 8.8 bolts and grade 50 steel, design a suitable framed connection.

4.2 Given the following information, design a suitable end plate connection using grade 43 steel, and grade 4.6 bolts. Factored end reaction 100 kN. Beam – 356 × 171UB45, Column – 203 × 203UC71, 180 kN end reaction (ultimate).

4.3 A 457 × 191UB74 is to be connected to a 254 × 254UC73, using a seated beam connection. The ultimate reaction from the beam is 120 kN. Using grade 50 steel and grade 8.8 bolts, design a suitable connection.

4.4 A stiffened seat bracket is to be designed to support a crane gantry girder, whose centreline is eccentric from the column centreline by 210 mm. The girder size is 457 × 191UB74, and the column size 203 × 203UC71. The ultimate end reaction is 130 kN. The beam web is at 90° to the stiffener. Use grade 50 steel, and E51 electrodes.

4.5 An eccentrically loaded connection (moment in the plane of the connection) is to be bolted to an existing column. The design load, $P = 250$ kN; eccentricity $= 300$ mm. Configuration given in sketch below. Design the

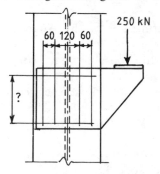

connection. (From *J. Institution of Structural Engineers*, volume **59A**, p. 207 – by courtesy of the Institution.)

4.6 A web splice as shown in the sketch below is proposed. Design the connection, using the minimum number of fasteners and a plastic basis of design.
Design load *P* (factored) = 800 kN.
Clear depth between flange fillets = 360 mm.
Web clearance = 4 mm.
Minimum fastener spacing = 60 mm.
Minimum edge distance of fasteners = 30 mm.
Design strength of fasteners = 130 kN.
(From *J. Institution of Structural Engineers*, volume **59A**, p. 207 – by courtesy of the Institution.)

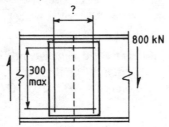

4.7 Design a beam to column connection given the data below:
Beam to column moment = 240 kNm (ultimate)
Beam to column shear = 300 kN (ultimate)
Beam – 356 × 171UB57
Column – 254 × 254UC73
Maximum haunch depth = 300 mm
Use grade 50 steel and HSFG bolts.

Solution to Question No. 4.5

Reproduced by kind permission of the Council, Institution of Structural Engineers.
Objective: To determine the most compact configuration on the basis of plastic behaviour.
An initial estimate of the number of fasteners can be obtained by assuming $a/b = 1$.
The group shown in the sketch is best represented as two columns spaced 180 mm apart.
From Table 4.7, $\Sigma r/na = 0.61$ approx.
and $e/(\Sigma r/n) = 300/(0.61 \times 180) = 2.7$

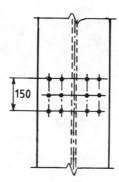

From Table 4.3, $P/nF = 0.32$ (by interpolation)
Thus $n = P/F \times 0.32 = 250/(0.32 \times 65)$
$\qquad\qquad = 12$ approx.
where F is the design shear strength of the fastener.
Try three rows at 90 mm vertical spacing.
Using Table 4.7 as before, $a/b = 180/180 = 1$
$n_r = 3,\ n_c = 2$
Thus $\Sigma r/nb = 0.64$ from Table 4.7
$e/(\Sigma r/n) = 300/(0.64 \times 180) = 2.6$
From Table 4.3, $P/nF = 0.33$
i.e. $P = 0.33 \times 12 \times 65 = 257.4$ kN
Repeating the above procedure for a modified vertical spacing of 75 mm (i.e.
$a/b = 1.2$), a value of $P = 249.6$ kN is obtained, which would be acceptable. The
exact solution for this configuration is $P = 248.0$ kN, indicating an accuracy of
within 1% for this tabular method.

Solution to Question No. 4.6

Reproduced by kind permission of the Council, Institution of Structural
Engineers.
Objective: To minimize the number of fasteners, using a plastic basis for design.
Maximum depth, b, of group $= 360 - (2 \times 30) = 300$ mm to bolt centres.
Minimum number of fasteners to resist P, ignoring eccentricity,
$\quad = 800/130 = 7$
Maximum number of fasteners per column
$\quad = (300/60) + 1 = 6$
Try $n_c = 2$, that is a/b ratio $= 60/300 = 0.2$
and $e = (0.5 \times 60) + 2 + 30 = 62$ mm
Assuming 10 fasteners, Table 4.7 gives $(\Sigma r/n)/b = 0.34$
$\Sigma r/n = [(\Sigma r/n)/b] \times b = 0.34 \times 300 = 106$
and $e/(\Sigma r/n) = 62/106 = 0.59$
Thus, from Table 4.3, $P/nF = 0.87$ (interpolating values)
and $n = 800/(0.87 \times 130) = 8$ (i.e., $n_r = 4$)

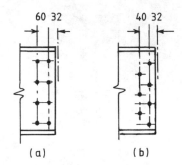

(a) (b)

$(\Sigma r/n)/b$ was derived conservatively from $n_r = 5$, but it may be shown that the result is effectively the same if $n_r = 4$ is used.

The tables are applicable to staggered groups, with little loss of accuracy, and the effect of reduced eccentricity for a more compact group of seven fasteners may be explored.

Using the staggered arrangement shown in part (b) of the sketch above, with column spacing 40 mm,

$a/b = 40/300 = 0.13$

$e = (40 \times 3/7) + 30 + 2 = 49 \text{ mm}$

From Table 4.7, $(\Sigma r/n)/b = 0.35$ (conservatively)

$e/(\Sigma r/n) = e/[(\Sigma r/n)/b]b = 49/(0.35 \times 300) = 0.47$

From Table 4.3, $P/nF = 0.92$, hence $P = 0.92 \times 7 \times 130 = 837.2 \text{ kN} > 800 \text{ kN}$

The exact solution for this configuration is $P = 816.3 \text{ kN}$, which indicates an accuracy better than 3% for this tabular method, notwithstanding the irregular fastener distribution.

REFERENCES

1. Fisher, J. W. and Struik, J. H. A. (1974) *Guide to Design Criteria for Bolted and Riveted Joints.* Wiley, New York.
2. McGuire, W. (1968) *Steel Structures.* Prentice-Hall, Englewood Cliffs, N.J.
3. American Institute of Steel Construction (1971) *Structural Steel Detailing.* 2nd Edn, AISC.
4. Packer, J. A. and Morris, L. J. (1977) A limit state design method for the tension region of bolted beam–column connections. *Struct. Engineer*, October.
5. American Welding Society (1983) Structural Welding Code. Publ. D1.1. AWS.
6. International Standards Organisation (1967) *Strength of Fillet Welds.* ISO Recommendations R617, Geneva.
7. European Convention for Constructional Steelwork (1976) *Recommendations for Steel Constructions.* ECCS.
8. Owens, G. W., Dowling, P. J. and Algar, R. J. (1976) *Bearing Stresses in Connections using Grade 8.8 Bolts.* Ceslic Report BC3 Eng. Structs. Laboratories, Imperial College, London.

Plastic design 5

5.1 INTRODUCTION

The plastic analysis method was validated from research undertaken at Cambridge by Professor Baker and his colleagues, and has been increasingly used for the design of structures such as portal frames. Generally, plastic design results in lighter structures than those designed by the elastic design method.

The basic behaviour of steel under direct tension is illustrated in Fig. 5.1 where stress and strain are proportional to each other up to the yield point, after which strain increases with only marginal change in stress. This ability of steel to stretch in a ductile fashion and yet still carry load is used in plastic design. The main features of the theory underlying the plastic method are:

(1) Failure of the entire cross-section of a member where a plastic hinge is formed;
(2) Consideration of the collapse of the structure by the formation of one or more plastic hinges, thus creating a mechanism.

The structure is checked for suitability against collapse (using factored loading). Under actual (service) loads, the structure behaves elastically and a check must be made to ensure that allowable deflections are not exceeded, using the elastic theory.

The plastic theory makes the assumption that the ultimate load is reached before secondary effects, such as member instability, cause failure. To ensure that local member instability does not occur in I sections, the flange width to thickness ratio is specified in BS5950. If the ratio exceeds the specified limit, then the section concerned would not be suitable for use in plastic design.

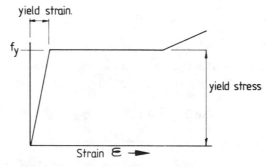

Fig. 5.1 Idealized stress/strain graph for steel.

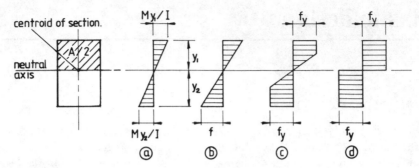

Fig. 5.2 Plastic behaviour of a homogeneous beam.

5.2 PLASTIC BEHAVIOUR OF A HOMOGENEOUS BEAM

Let a beam with an axis of symmetry (Fig. 5.2) be subjected to bending in the plane of symmetry.

(1) A small bending moment is applied and the extreme fibres are stressed below yield stress (Fig. 5.2a).
(2) The moment is increased until yield stress is attained in the extreme fibres, (Fig. 5.2b) and, with a further increase, yield stress spreads from outer fibres towards the neutral axis (Fig. 5.2c).
(3) With a continuing increase in moment yield spreads from the outer fibres towards the neutral axis until the tensile and compressive zones meet (Fig. 5.2d) and the section becomes fully plastic.

The magnitude of the plastic moment M_p in stage (d) can be calculated in terms of the yield stress σ_y. The resultant tension and compression are both equal to $A\sigma_y/2$, which when multiplied by the distance between the centroids of the compression and tensile areas, equals the plastic moment M_p. i.e. $M_p = (A\sigma_y/2) \times$ lever arm.

5.3 PLASTIC MODULUS

The plastic modulus S is the algebraic sum of the first moments of area about the equal area axis (for symmetrical sections the equal area axis coincides with the neutral axis).

For the I section shown in Fig. 5.3 the plastic modulus is:

$$2bcy + 0.5dt \times 0.25d = 2bcy + 0.125d^2t$$

At full plasticity the maximum moment a section can sustain is equal to Sf_y where $f_y =$ yield stress.

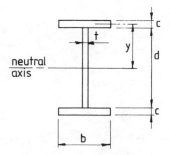

Fig. 5.3 Symmetrical section.

5.4 SHAPE FACTOR

The shape factor is a ratio of plastic modulus S to elastic modulus Z. Therefore shape factor = S/Z and is dependent only on the shape of the cross-section considered. For a rectangular section of breadth b and depth d

$$Z = bd^2/6 \quad \text{and} \quad S = bd^2/4 \quad \text{and the shape factor} = 6bd^2/4bd^2 = 1.5$$

For a solid circular section the shape factor is 1.7, and for an I section it is approximately 1.15.

5.5 SINGLE SPAN BEAMS

The distribution of elastic and plastic stress along a simply supported beam is shown in Fig. 5.4. At collapse load the zone below the point load is fully plastic, and is transformed into a 'plastic hinge'. The beam each side of the plastic zone remains elastic.

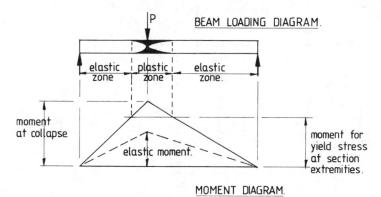

Fig. 5.4 Simply supported beam.

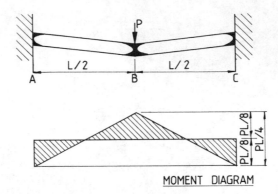

MOMENT DIAGRAM

Fig. 5.5 Fixed ended beam.

A beam which is fixed at one end, requires two plastic hinges to form before collapse (Fig. 5.6), and if fixed at both ends requires three plastic hinges for collapse to occur (Fig. 5.5).

Example 5.1

Consider a fixed end beam span L with a factored point load P in centre of span (Fig. 5.5). Three plastic hinges are required for collapse (one each end and one under the load). The positions of the maximum moments occur at the plastic hinges. Make maximum moments equal in magnitude.

Thus, free moment at B = $PL/4$
and moment at A and C = $PL/4 \div 2 = PL/8$

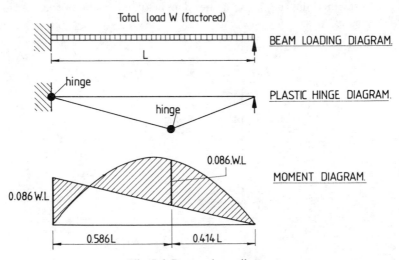

Fig. 5.6 Propped cantilever.

Note: If there are a number of loading conditions, a separate calculation for each is required. The *maximum* plastic moment derived is used to select a beam section.

The plastic moments obtained from different loading conditions must not be combined.

Example 5.2

A propped cantilever, span L, with factored load W uniformly distributed is shown in Fig. 5.6.

By equating the span moment to the support moment the value and position of each can be found using the 'free' bending moment diagram. The plastic moment for a propped cantilever equals 0.686 × maximum free moment,

$$M_p = 0.686WL/8 = 0.086WL \quad \text{or} \quad WL/11.66.$$

This moment occurs at the fixed end and at 0.586L from the fixed end.

If the position of the span hinge is approximated to be at midspan and it is assumed that $M_p = WL/12$, the error is only 3%.

5.6 CONTINUOUS BEAMS

Consider a continuous beam having equal spans L, and a UDL (factored) of w per unit length (Fig. 5.7).

Hinges will form at midspan of each span and at each support. Therefore the magnitude of plastic moment will be half that of the free moment,

$$M_p = 0.5WL^2/8 = WL^2/16$$

If the illustration in Fig. 5.7 had pinned supports at A and E, and the beam was of uniform cross section, then the plastic moment for the end spans AB and DE would be greater than for the internal spans. In this case the end spans alone would collapse, and are therefore the critical spans. The magnitude of the plastic moment for the end spans would be equal to that stated in the example of the propped cantilever (Fig. 5.6).

5.7 UNEQUAL SPANS

Consider the continuous beam with uniform cross-section shown in Fig. 5.8 with 'factored' loads. The value of the plastic moment of the *critical* span (the span that collapses first) will be used to design the beam.

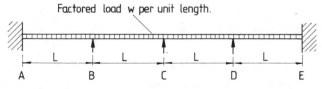

Fig. 5.7 Continuous beam with equal spans and fixed ends.

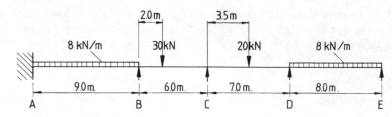

Fig. 5.8 Continuous beam with unequal spans.

The magnitude of the maximum free bending moments for the four spans are:

AB: $8 \times 9^2/8 = 81\,\text{kNm}$ BC: $30 \times 4 \times 2/6 = 40\,\text{kNm}$

CD: $20 \times 7/4 = 35\,\text{kNm}$ DE: $8 \times 8^2/8 = 64\,\text{kNm}$

Span AB has the largest moment, but DE develops 2 hinges, as against 3 hinges for span AB. Spans AB and DE must be investigated.

Span AB: $M_p = 0.5 \times 81 = 40.5\,\text{kNm}$

Span DE: $M_p = 8 \times 8^2 \times 0.86 = 44.0\,\text{kNm}$ (larger M_p used to calculate beam section).

S reqd. $= 44 \times 10^3/275 = 183.3\,\text{cm}^3$

Shear to right of $D = 8 \times 4 + 44/8 = 37.5\,\text{kN}$ (governs)

Shear to left of $D = 20/2 +$ moment diff. (C–D)$/7$ (assume negligible) $= 10\,\text{kN}$

Try $203 \times 133\,\text{UB}\,25$ grade 43.

Shear capacity $P_v = 0.6 p_y tD$ (code 4.2.3)

If the shear force exceeds $0.6 P_v$ the moment capacity is reduced

(code 4.2.5, 4.2.6)

Here $P_v = 0.6 \times 275\,tD = 99\,tD = 99 \times 5.8 \times 203.2/10^3 = 116.7\,\text{kN}$

Actual shear force $= 37.5\,\text{kN}$ so no reduction in moment.

5.8 EFFECT OF SHEAR ON PLASTIC MOMENT CAPACITY

If shear occurs across a beam section where the bending moment is high, the plastic moment value at the hinge position may be reduced. The value of the reduction depends upon the type of beam section used, i.e. compact or semi-compact. For reduction formulae refer to BS5950 4.2.6 (see also Chapter 2, Section 2.2.4).

Generally, the influence of shear is negligible, but there are conditions where a significant reduction of plastic moment can occur.

5.9 EFFECT OF AXIAL FORCE ON PLASTIC MOMENT CAPACITY

Before considering the design of portal frames by the plastic theory, we shall consider the effect of axial force (compression or tension) in combination with plastic moment.

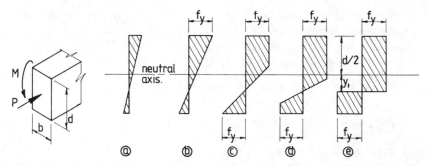

Fig. 5.9 Effect of axial force on plastic moment capacity.

Axial force tends to reduce the value of the plastic moment. An expression is derived for the reduced M_p for a rectangular section subjected to an axial force. The member is considered to be restrained in order that buckling will not occur.

Consider a rectangular member of width b and depth d subjected to an axial compressive force P together with a moment M in the vertical plane (Fig. 5.9).

The values of M and P are increased at a constant value of M/P until the fully plastic stage is attained, and the values of M and P become:

$$M_{pa} = 0.25 f_y b (d^2 - 4y^2) \tag{1}$$
$$P = 2y_1 \times b f_y \tag{2}$$

where f_y = yield stress

$\quad y_1$ = distance from the neutral axis to the stress change for
$\qquad M_p$ without axial force, $M_p = f_y b d^2 / 4 \tag{3}$

If axial force acts alone $-P_y = f_y b d \tag{4}$
at the fully plastic state.

From equations (1) to (4) the interaction equation can be obtained:

$$M_x / M_p = 1 - P^2 / P_y \tag{5}$$

As an empirical rule, when axial load is less than 15% of the maximum that can be carried by the section, the effect on plastically designed members will be relatively small. Values of the reduced plastic modulus may be calculated from the formulae quoted in Chapter 2.

5.10 PLASTIC DESIGN RESTRICTIONS

(1) Plastic design should only be used when the number of loading cycles is small, and not in fatigue situations.

(2) Local buckling due to concentrated loads must be considered. Stiffeners must be provided if the concentrated load is more than the web buckling

or bearing capacity (see code 4.5.2.1 and 4.5.3) or if shear occurs (near a plastic hinge) which is greater than 10% of the shear capacity of the member, then a full depth stiffener must be provided within a distance of $D/2$ either side of the hinge location (code 5.3.6).

All stiffeners should be designed elastically (see Chapter 2).

5.11 PLASTIC DESIGN OF PLANE FRAMES

For a frame to transform into a mechanism and collapse, it is necessary for a number of hinges to develop. The number of hinges is equal to the number of redundant restraints plus one. A method of determining the number of hinges for overall collapse of any plane frame is illustrated in Fig. 5.10. The same rule applies to the beams previously considered, with the exception of the continuous beam, in which only one span reached the point of collapse and should be regarded as a partial collapse of the whole structure. In the case of complicated plane frames or multi-storey structures, partial collapse may govern the design. A typical case of partial collapse in a multi-storey structure occurs when hinges develop in the beams only, thus causing overall collapse but which is, from a mathematical point of view, partial collapse. The amount of calculation required to justify multi-storey frames designed plastically rises rapidly when structures over two storeys are considered. Also, the presence of significant axial load reduces the amount of bending moment a hinge can develop, thus making plastic design less applicable. It is for this reason that plastic design is generally limited to *complete structures* of four storeys or less with moderate member axial loads.

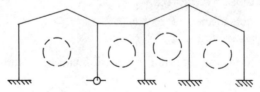

Number of redundancies = (4 rings × 3) –1 pin = 11 unknowns.
Number of hinges to cause collapse = 11 – 1   = 10

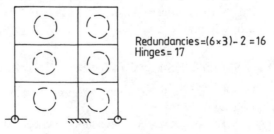

Redundancies = (6 × 3) – 2 = 16
Hinges = 17

Fig. 5.10

5.12 SINGLE BAY PORTAL FRAMES

One span portal frames are the most common application of plastic design. There are basically three methods of designing such frames which are:

(1) The virtual work method
(2) The graphical method
(3) The reactant moment diagram method.

These three methods are described in the following examples. When frames are designed by one of these methods a check may be carried out using one of the alternative methods.

5.13 VIRTUAL WORK METHOD

A single bay portal frame with 'pinned' column bases has one redundant restraint, and therefore two hinges must form to cause a collapse. The same frame with 'fixed' bases has three redundant restraints, and four hinges are required for collapse. Fig. 5.11 denotes alternative locations of the hinges for a fixed base portal frame. The alternative mechanisms are called collapse modes.

For the frame shown in Fig. 5.11 the collapse mode which will govern depends on the relative plastic moduli of the beam rafter and column, and the relative magnitude of the loads P and W. Each collapse mode requires separate calculation, and the least value of the loads causing collapse is the correct result and hence the true collapse mode.

Example 5.3

Determine the collapse moment for the rectangular portal frame shown in Fig. 5.12. Relative rotations and collapse modes are shown in the sketch.

CONSIDER MODE (a)
The load P_1 causes point B to move a distance of $h\phi = 10\phi$.
$\qquad$ P_2 causes point E to deflect a distance of $L\phi/2 = 9\phi$
Work done $= 10P_1\phi + 9P_2\phi = 80\phi + 108\phi = 188\phi$

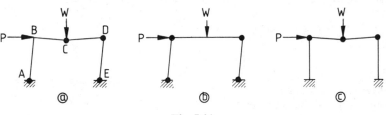

Fig. 5.11

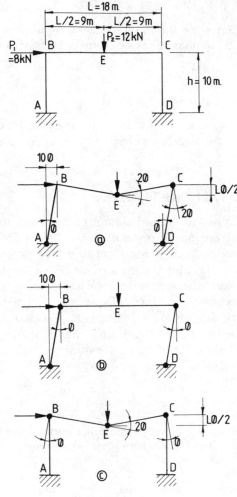

Fig. 5.12

At hinges the work done = sum of moments × angular rotation

$$= M_p\phi + 2M_p\phi + 2M_p\phi + M_p\phi \text{ at points A, B, C, D respectively}$$

$$= 6M_p\phi$$

Equating the above gives $6M_p\phi = 188\phi$

$M_p = 31.3\,\text{kNm}$

CONSIDER MODE (b)

All hinges rotate through the angle ϕ

$\therefore 10P_1\phi = 4M_p\phi$ and $M_p = 80\phi/4\phi = 20\,\text{kNm}$

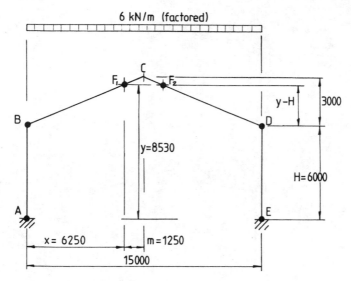

6 kN/m (factored)

$y-H$ 3000

$y=8530$

$H=6000$

$x=6250$ $m=1250$

15000

Fig. 5.13 Portal with fixed base and symmetrical vertical loading.

NOW CONSIDER MODE (c)
A rotation of ϕ at B and C gives a rotation at E of 2ϕ
$$\therefore 9 \times P_2\phi = 4M_p\phi$$
$$9 \times 12\phi/4\phi = M_p = 27\,\text{kNm}$$
Therefore, the maximum plastic moment of 31.3 kNm occurs in mode (a) and this moment is used to design the members.

Example 5.4. Design of a single bay pitched portal flame with symmetrical vertical loading

Data (as shown in Fig. 5.13):

Frame centres	5.0 m
Span of portal	15.0 m
Eaves height	6.0 m
Eaves to ridge height	3.0 m
Purlin spacing	1.25 m

Loading UDL 6 kN/m (factored).

The possible failure modes for a fixed base pitched portal frame are shown in Fig. 5.14, but as this frame carries only symmetrical vertical loading the hinges will form at locations shown in mode 1. Two hinges will form in the rafters at the nearest purlins to the roof apex, but take as one hinge (points F1, F2) due to symmetry of the frame and loads. Purlins are spaced at 1.25 m horizontally. The rotations are evaluated using the following expressions:

$$\phi AB = \phi DE = \phi; \; \phi BF_1 = \phi DF_2 = \phi H/(y-H); \; \phi F_1 = \phi F_2 = 0$$
$$\phi BF_1 = \phi DF_2 = 6\phi/(8.53-6) = 2.37\phi$$

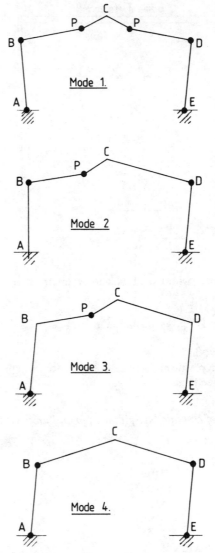

Fig. 5.14 Failure modes for fixed base portal frame.

The centroid of loading on BF_1 changes vertically a distance of $2.53 \times 2.37\phi = 6\phi\,\text{m}$ approximately.

$$\text{Work done} = 2(6 \times 6.25 \times 6\phi) = 450\phi\,\text{kNm}$$

The load on F_1 to F_2 moves a vertical distance of $6.25 \times 2.37\phi = 14.8\phi$

Work done $= 6 \times 2.5 \times 14.8\phi = 222\phi\,\text{kNm}$

Total work done by loads $= \Sigma W_\Delta = 450\phi + 222\phi = 672\phi\,\text{kNm}$ \hfill (1)

Rotations in hinges $= \phi A = 0$

$\phi B = \phi D = \phi AB + \phi BF = \phi + 2.37\phi = 4.37\phi$

$\phi F_1 = \phi F_2 = \phi BF_1 + \phi F_1 F_2 = 2.37\phi + 0 = 2.37\phi$

$\therefore M\phi = 2M_p(\phi + 4.37\phi + 2.37\phi) = 15.48 M_p \phi$ (2)

Equating (1) and (2) $672\phi = 15.48 M_p \phi$

whence $M_p = 43.4\,\text{kNm}$

5.14 GRAPHICAL METHOD

The advantage of this method lies in its direct visual verification of the plastic moment diagram. In addition, it offers the possibility of easy adjustment of haunches on the plastic moment diagram to obtain smaller rafter plastic moments. Since the method does offer flexibility in arriving at a solution, the authors recommend that one span portals be analysed firstly by the graphical method, and then checked by the reactant moment diagram method.

A typical pitched roof portal analysed by the graphical method follows.

Example 5.5 M_p for a portal frame with electrically operated travelling crane

Find M_p (the plastic moment) for a portal frame with an electric overhead crane, as illustrated in Fig. 5.15, solving by drawing.

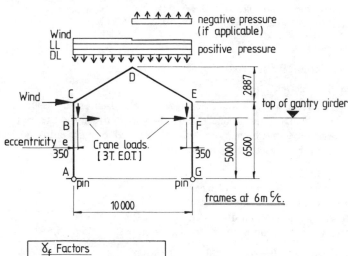

γ_f Factors				
	LL	DL	Crane	Wind
Case 1	1.2	1.2	1.2	1.2
Case 2	1.6	1.4	1.6	—
Case 3	—	1.0	—	1.4

(γ_f from BS 5950)

Fig. 5.15 Summary of typical loads and dimensions for Example 5.5.

The roof pitch is 30°. Neglect the effect of the wind acting vertically on the roof. Horizontal wind pressure = $1\,kN/m^2$.

To obtain the free bending moment diagram from the portal frame, put point 'G' on an imaginary roller, with point 'A' remaining as a pin.

PRELIMINARY CALCULATIONS
(1) Forces due to dead load and live load on roof

Vertical loading on walls		Vertical loading on roof	
	kN/m^2		kN/m^2
Cladding	0.14	Superimposed:	0.75
		Dead	
Self weight of rails	0.03	Cladding:	0.14
		Purlins:	0.03
		Rafters:	0.23
	0.17		0.40

Purlins and cladding rails are cold formed sections; the superimposed loading is that given in BS6399: Part 1 which has superseded CP3: Chapter V: Part 1.

Moment due to dead load and live load on roof (unfactored) (Fig. 5.16)
Moment due to roof dead load = $0.40 \times 6 \times 10^2/8 = 30.0\,kNm$
Moment due to roof live load = $0.75 \times 6 \times 10^2/8 = 56.3\,kNm$

(2) Crane loading

3 ton capacity crane, 9.3 m span.

Horizontal crane loading
This may be shared between each side of the portal, based on the assumption that the crane wheels are flanged, and in effect share the load between the two rails. Check that the crane wheels are flanged when the vendor is selected, or place entire horizontal crane load at point B for a more onerous case.

Vertical crane loading
Maximum wheel load = $26.5\,kN$ (2 wheels)
Minimum wheel load = $7.25\,kN$ (2 wheels)
Maximum reaction at column due to crane = $2 \times 26.5 = 53\,kN$
Minimum reaction at column due to loaded crane = $2 \times 7.25 = 14.5\,kN$

Moment due to vertical crane loading (unfactored)
Moment at B = $53 \times 0.35 = 18.55\,kNm$
Moment at F = $14.5 \times 0.35 = 5.08\,kNm$
Load on crane bracket is 350 mm eccentric from column centreline.

Transverse crane loading (see BS6399: Part 1)
Transverse load due to crab and load = $0.1(6.0 + 30) = 3.6\,kN$
Shared between points B and F, i.e. 1.8 kN each.

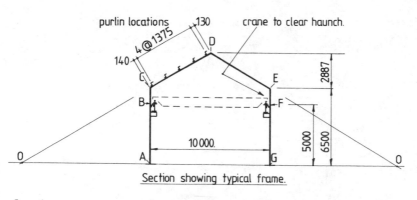

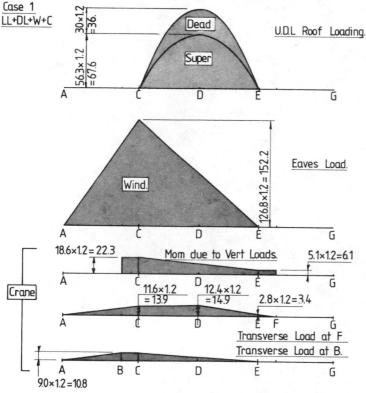

Fig. 5.16 Free bending moment diagrams.

Moment due to transverse crane loading
Moment at $B = 1.8 \times 5 = 9.0\,kNm$
The moment caused by transverse crane loading at F is illustrated in Fig. 5.17. Project a line from point F to point A. Moments on the frame are equal to the load of 2 kN multiplied by the distance the frame lies away from line AF.

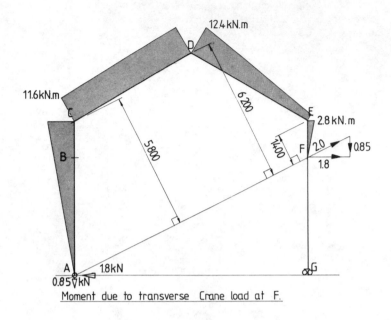

Moment due to transverse Crane load at F.

Moments shown on
outside of frame for clarity

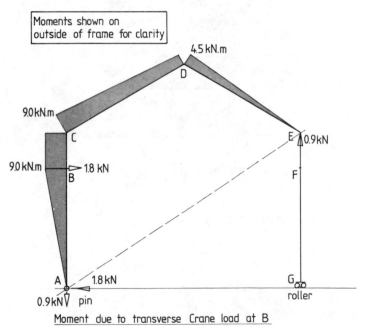

Moment due to transverse Crane load at B

Fig. 5.17 Moments due to transverse crane loads.

Wind loading
Assume entire wind force on side of building is taken at eaves.
Force at eaves $= 1 \times 6 \times 6.5/2$
$$= 19.5 \, \text{kN}$$

Moment due to horizontal wind load (unfactored)
Moment at eaves $= 19.5 \times 6.5 = 126.8 \, \text{kNm}$.

At the top of Fig. 5.16 the frame is drawn to scale. Positions CE, DE are projected down to the baseline at points 'O'.

Free moment diagrams are drawn below, suitably factored. Points C, D, E

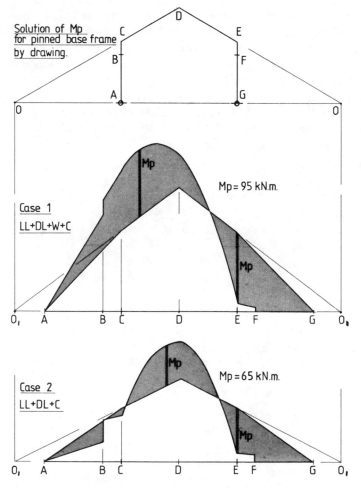

Fig. 5.18 Bending moment envelopes (developed from Figs 5.16 and 5.17).

on the free moment diagrams are drawn vertically in line with points C, D, E on the frame. Lines AC and EG on the free moment diagrams are drawn to the same scale as the frame above.

These free moment diagrams are then summed, as shown on Fig. 5.18. The plastic moment M_p is found by projecting lines from points 'O' through the bending moment envelopes such that equal maximum values of M_p are obtained. These values may be checked by calculation, and have in this case been found to lie within 5% of the values obtained by drawing (see Example 5.6), which verifies the method and indicates the accuracy possible by a drawing solution.

5.15 REACTANT MOMENT DIAGRAM METHOD

This method of solution has the advantage that it can readily be programmed into a desk computer. The basic considerations are illustrated in Fig. 5.19. Free moment diagrams are sketched and moment values obtained for points A to G. The redundant reactant moment diagram counteracts the applied moments. The summation of both diagrams yields the value of the plastic moment M_p, and is written algebraically in the equations for fixed bases. If a base is pinned the moment value must be set to zero. These equations may then be solved either simultaneously, or by matrix, preferably using a small computer.

Example 5.6

Check the plastic moment obtained by drawing in the previous example.

Determine free moments
Recalculate all moments on the basis of the frame being split at the apex, treating it as two cantilevers.
Moment at:
A = 291.8 kNm
Purlin point 3 = 26.5 kNm (*see below*)
E = 106.5 kNm
G = 123.4 kNm

For solution by calculation, let purlins be numbered 1 to 'n' from apex to roof. Put moment for each purlin point into the reactant moment diagram equations and solve for successive purlin points. The largest value of M_p found by this method is the design case.

For this design the equations for M_p (Fig. 5.19) are:

At A: $291.8 - m - 9.387R + 5S = 0$
At point 3: $26.5 - m - 1.44R + 2.495S = -M_p$
At E: $106.5 - m - 2.887R - 5S = +M_p$
At G: $123.4 - m - 9.387R - 5S = 0$

Note: For the collapse mode adjacent plastic moments have opposite signs.

Using the method and equations illustrated in Fig. 5.19 these equations can

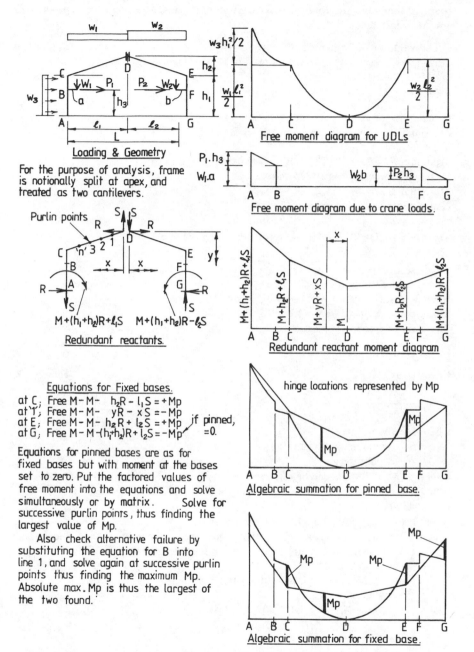

Loading & Geometry

Free moment diagram for UDLs

For the purpose of analysis, frame is notionally split at apex, and treated as two cantilevers.

Free moment diagram due to crane loads.

Purlin points

Redundant reactants.

$M+(h_1+h_2)R+l_1S$ $M+(h_1+h_2)R-l_2S$

Redundant reactant moment diagram

Equations for Fixed bases.

at C; Free M – M – $h_2R - l_1S$ = +Mp
at 'I'; Free M – M – $yR - xS$ = –Mp
at E; Free M – M – $h_2R + l_2S$ = +Mp   if pinned,
at G; Free M – M – $(h_1+h_2)R+l_2S$ = –Mp   = 0.

Equations for pinned bases are as for fixed bases but with moment at the bases set to zero. Put the factored values of free moment into the equations and solve simultaneously or by matrix.   Solve for successive purlin points, thus finding the largest value of Mp.

Also check alternative failure by substituting the equation for B into line 1, and solve again at successive purlin points thus finding the maximum Mp. Absolute max. Mp is thus the largest of the two found.

hinge locations represented by Mp

Algebraic summation for pinned base.

Algebraic summation for fixed base.

Fig. 5.19 Summary of the reactant moment diagram method for a portal frame.

be solved simultaneously (or by matrix) to give $R = 16.6\,\text{kN}$, $M = 51.7\,\text{kNm}$, $S = -16.8\,\text{kN}$ and $M_p = 91.0\,\text{kNm}$.

The value of M_p obtained by calculation agrees within 5% of the value found by drawing which can be considered a reasonable check on the calculated value.

Example 5.7

Calculation for column and rafter sizes in grade 43 steel using the maximum value for M_p found from the analyses in Examples 5.5 and 5.6, i.e. $M_p = 95\,\text{kNm}$.
$M_p = 95\,\text{kNm}$
Minimum $S_x = 95 \times 10^6/275 \times 10^3 = 345.5\,\text{cm}^3$
Distance between purlins: 1375 mm
Minimum $L/r_y = 70$
Hence minimum r_y for rafter $= 1375/70$
$$= 19.6\,\text{mm}$$

Try $305 \times 102\text{UB}28$ $(r_y = 20.8\,\text{mm})$ for rafter and column

Check section proportions:
for a plastic section $b/T = 8.5\varepsilon$ (code Table 7)
where $\varepsilon = \sqrt{(275/p_y)} = 1$ when $p_y = 275\,\text{N/mm}^2$
Rafter/column $b/t = 51/8.9 = 5.7 < 8.5$
Therefore the section is plastic

Check restraints *after* deflection

Check sway stability   (code 5.5.3.2)

$$\frac{L}{D} \not> \frac{44L(\rho)}{\Omega h[4 + (\rho_1 L_r/L)]}\frac{275}{p_{yr}}$$

where L = span of bay
       D = minimum depth of rafter
       h = column height
$\rho = (2I_c/I_r)(L/h)$ for a single bay frame $= 2 \times 10/6.5 = 3.077$
(since $I_c = I_r$ = minimum second moment of area of column and rafter).
Ω = arching ratio, W_r/W_o
where W_1 = factored vertical load on the rafter
   and W_o = maximum value of W_r that could be carried by a fixed ended beam
              of span L (10 metres) with same section as the rafter
              $= 8p_yS_x/L \times 10^3$
              $= 8 \times 275 \times 407.2 \times 10^4/10^4 \times 10^3$
              $= 780\,\text{kN}$
   $W_r = 105.6\,\text{kN}$ (DL $0.40 \times 6 \times 1.4$)
                              (LL $0.75 \times 6 \times 1.6$)
   $\therefore$ $\Omega = W_r/W_o = 105.6/780 = 0.135$
   and L_r = total developed rafter length $= 2 \times 5774$
                                           $= 11548\,\text{mm}$
       P_{yr} = design strength of rafter $= 275\,\text{N/mm}^2$

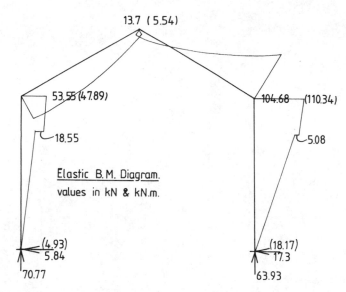

Fig. 5.20 Elastic bending moment diagram. (Values in kN and kNm.)

Then $\dfrac{L}{D} = \dfrac{10^4}{308.9} = 32.373 \not> \dfrac{44}{0.135} \times \dfrac{10^4}{6500} \times \dfrac{3.077}{4 + (3.077 \times 11.548/10)} \times \dfrac{275}{275}$

$$32.373 \not> 222.5$$

Therefore the frame is satisfactory for sway stability, although deflection should be checked.

Check frame deflection (Fig. 5.20)
Using the area moment method, the horizontal deflection at the eaves and crane bracket level, and the vertical deflection at the ridge, may be found using standard formulae (such as those given in the Appendix). Working from first principles, the elastic moment diagram may be determined for the frame as shown in Fig. 5.21 and deflections found as follows (Fig. 5.21).
In general, $\Delta = \Sigma Ay/EI$
Where $E = 205 \times 10^3 \, \text{N/mm}^2$ and $I = 5421 \, \text{cm}^4$ for the section selected.
    Δ = deflection.
At E $\Delta_E = 10^{12}[(90.85 \times 5/2)3.333 + (95.93 \times 1.5)5.75 + (14.41 \times 1.5/2)6]/EI$
        = 148 mm to the right ($= h/44$)
Deflection at this point exceeds a notional allowable of $h/200$, therefore change column size and recalculate elastic moments and deflections. (Crane consideration governs.)
If this frame did not carry an EOT crane, this extra check on deflection (in addition to code 5.5.3.2 requirement) would not be necessary.

Recalculating deflection – see revised moment diagram, Fig. 5.20.
Note: Values in brackets are initial bending moments; final bending moments are shown alongside.

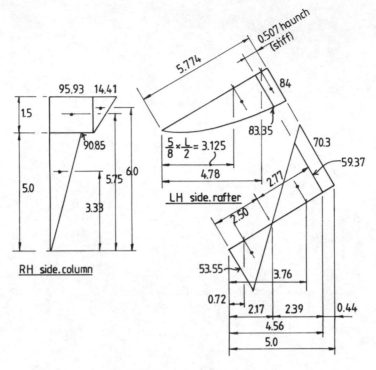

Fig. 5.21 Area moment method.

New I value for column $= (200/46) \times I$ of 5421×1.08
$$= 25\,464\,\text{cm}^4 \text{ (increase in } I \text{ value assumed in this case attracts}$$
an extra 8% moment)

Try $457 \times 152\,\text{UB}60$ for column only

Note: Increasing the column size does NOT increase the strength of the whole frame and the ultimate strength at collapse is the same.

Check section proportions:
$b/T = 71.5/13.3 = 5.38 < 8.5\varepsilon$, shape acceptable   (code Table 7)
$\Delta_E = 10^{12}[(86.5 \times 5/2)3.333 + (91.58 \times 1.5)5.75 + (13.1 \times 1.5/2)6]/EI$
$\quad = 30.0\,\text{mm}$ to the right, i.e. $h/216$ – acceptable.

Nett deflection at crane rail level
$\Delta_B - \Delta_F = 10^{12}\,(17.3 - 5.84)5^3/6EI = 3.0\,\text{mm}$ nett relative movement.
Check with crane manufacturers that this figure is acceptable when the vendor is selected.

Check deflection at centre
Let the total depth of the ridge haunch (see Fig. 5.22)
$= (\text{rafter depth}/\cos\theta) + \text{rafter depth} - \text{flange thickness} - \text{root radius}$
$= (308.9/\cos 30°) + 308.9 - 8.9 - 7.6 = 356.6 + 292.4 \simeq 650\,\text{mm}$

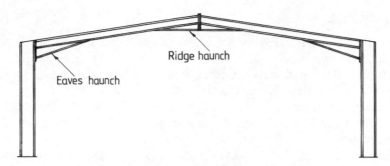

Fig. 5.22 Portal frame haunches.

Length of ridge haunch from centre $= 292.4/\tan\theta$
$$= 507\,\text{mm}$$

Consider this central haunch as relatively 'stiff'; then deflection at centre may be calculated as follows:
$$\Delta_\text{D} = 10^{12}[(5.774 \times 84 \times 0.666)3.125 - (0.507 \times 83.7)4.78$$
$$+ (53.55 \times 2.50/2)0.72 - (59.37 \times 2.77/2)3.76]/EI$$
where $I = 5421\,\text{cm}^4$
$$= 49\,\text{mm, i.e. } L/204 - \text{acceptable}$$

Column and rafter sizes are now shown to be adequate for plastic moment and service deflection.

Check torsional restraint of column and rafter
Consider rafter, 305 × 102UB28.
Let depth of haunch/depth of rafter = 1 (same section)
Maximum distance between restraints to the compression flange is given by
$$L_\text{t} = K_1 r_y x/\sqrt{(72x^2 - 10^4)} \qquad\qquad\qquad \text{(code 5.5.3.5)}$$
$K_1 = 620$ and $x = D/t$
Then $L_\text{t} = 620r_y x/\sqrt{(72x^2 - 10^4)}$
$$= 620 \times 20.8 \times 34.71/\sqrt{(72 \times 34.71^2 - 10^4)}$$
$$= 1615\,\text{mm} > 1375 \text{ provided, which is acceptable.}$$

Provide torsional restraint bracing at purlin locations Nos. 5 and 4 on each side of the roof (i.e., at eaves and adjacent purlin).
Note: The designer may find that a heavier rafter section may eliminate the need for torsional restraint 'knee' bracing thus reducing fabrication costs.

At this stage a check of dimension L_m (code 5.3.5) is generally required, but as the column is obviously in the elastic range, being much larger than the previous 'uniform' size that was retained for the rafter, then torsional restraint is required at the eaves level only. Check column satisfies conditions noted in code 4.8.3. Note column L_e is described in code Figs 18 and 19.

Provide torsional restraint bracing at the eaves cladding rail only. Spacing of side cladding rails is generally dependent on the most suitable regular spacing for the cladding and the particular rails selected. Optimum spacing of rails $= 1600\,\text{mm}$. Spacing selected $= (6500 - 300)/4 = 1550\,\text{mm} < 1600\,\text{mm}$.

5.16 PORTAL FRAME HAUNCHES

The eaves haunch is often fabricated from the same section as the rafter (Fig. 5.22), but it is good practice to use the same section size, one serial weight heavier. This haunch reduces the local bending stress and increases the lever arm of the bolts connecting the rafter to the column.

A haunch is also generally placed at the ridge of a frame to reduce deflection. As the plastic hinges develop outside the haunch length, the addition of a haunch in this location has no effect on the strength of the rafter.

When the top flange of the rafter (in the region of the eaves) is in tension, instability is prevented by providing torsional restraint at the eaves at a maximum distance L_t from the first restraint near the point of contraflexure. The length of the haunch b is usually extended to coincide with a rafter moment of 0.87 M_p, i.e. to a point where the rafter behaves elastically.

The entire haunch should thus lie within the elastic range, with M_p occurring in the column, in the horizontal plane just below the haunch flange (with crane loading within the column length).

The bolted interface between haunched rafter and column may be designed elastically (see Chapter 4, beam to column connections) using the plastic moment at that point determined by analysis.

For the length of haunches for pinned or fixed base portals approximate formulae are given in Fig. 5.23 (these formulae apply when vertical loading governs).

An exact formula is stated for a pinned base portal where, for a given haunch size, the largest M_p is found by examining various plastic hinge locations.

The term x_1 is measured horizontally from the frame centre to the assumed plastic hinge (Fig. 5.23).

The rafter and haunch should be checked for elastic lateral stability within

Mp fixed base $= (1 - a/h_1) Mp'$

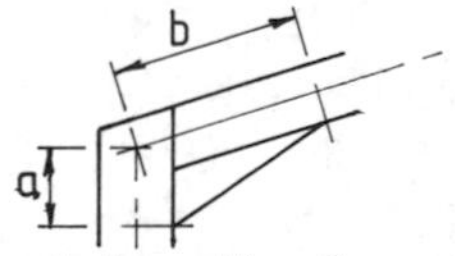

Exact formula for reduction in Mp due to haunching, for a pinned base (where symmetrical vertical UDL governs)

$$Mp = Mp' \times \left[\frac{2h_1{}^2 + h_1 h_2 - 2ah_1 - ah_2 - (h_2/x_1)(h_1/a)}{h_1(2h_1 + h_2 - a) - (h_2/x_1)(h_1)} \right]$$

if terms in x_1 are omitted, then Mp virtually at apex,

$$Mp = \left[1 - \left(\frac{h_1 + h_2}{h_1}\right) \left(\frac{a}{2h_1 + h_2 - a}\right) \right] \times Mp'$$

with acknowledgement to Gordon Stamper

Fig. 5.23 Approximate formulae for reduction in M_p due to haunching (symmetrical vertical UDL on roof).

the dimension L_t (code 5.5.3.5). Another procedure detailed by Morris [11] is based on the assumption that the top flange of the rafter is restrained by purlins, neglecting the effect of the central (third) flange created by adding a haunch. Utilizing the moment diagram obtained from a plastic analysis, Morris evaluates the moments at five points (four equal intervals) between column centreline and the point of contraflexure. Elastic moduli for the rafter and haunch at these points are found and maximum stresses at the five points f_1 to f_5 are calculated, adding an allowance for axial stress of $0.6Pd$ where P is the factored axial load in the rafter and d is the distance between the flange centroids of the rafter. The term $0.6Pd$ can be neglected where axial stress is less than $15\,\text{N/mm}^2$. The value of factor k is given by:

$$k = \frac{1}{12p_y}[f_1 + 3f_2 + 4f_3 + 3f_4 + f_5 + 2(f_s\,\text{max.} - f_e\,\text{max.})]$$

where f_s max. is the largest positive stress within the range considered and f_e max. is the largest possible value of f_1 or f_2. Moments are considered positive in this case where they produce compression in the unrestrained flange. Positive values of f_1 to f_5 only are used, negative values are set to zero. The research of Horne and Ajmani then shows (for a two-flanged member) that:

$$(p_y - f_s\,\text{max}) > \frac{2p_y\sqrt{k} \times (L/100\,r_y)}{20 - k \times 0.065p_y - 3000\,T^2/D^2(L/100r_y)^2}$$

where D = overall depth of the haunch and T is the thickness of the rafter flange. If the right hand side of the above inequality is negative, the member is not stable and an additional lateral restraint must be added and the new shorter distance checked for stability. A 'heavier' haunch section with a thicker flange could be alternatively tried in which case the modified stability condition becomes:

$$(p_y - f_s\,\text{max.}) >$$

$$\frac{2p_y\sqrt{k} \times (L/100r_y)}{20 - k\left[0.065p_y - 3000\left(\dfrac{\text{rafter } T}{D}\right)^2 + 1200 \times \left(\dfrac{\text{haunch } T}{D_1}\right)^2 (L/100r_y)^2\right]}$$

where D_1 = Max. depth of haunch.
    D = minimum depth of haunch.

There is no need to check for haunch stability if:

$$D/T \text{ rafter} < 220/\sqrt{p_y}$$

The welds to the haunch flanges are generally specified as full strength butt welds, and the haunch web is fillet welded to the rafter. As a matter of academic interest, at the rafter end of the haunch, the component of the rafter flange force in the haunch flange generally lies in the region of 60–75%, the remainder of the force being carried by the haunch web.

5.17 CONNECTIONS AND DETAILS

The eaves haunch may be designed as outlined previously and the ridge haunch as in Section 3.7, splice connections. The baseplate and bolts should be designed to act elastically (even if the column itself has a full plastic moment), as noted in Section 3.5. Column brackets may be designed as noted in Section 4.4 with the crane gantry girder bolted to the bracket.

At haunch and crane bracket, when a plastic hinge forms in the column, stiffeners are required if:

$$d_c/t_c > 55\sqrt{(275/p_y\,\text{col.})}$$

where d_c = depth of column and t_c = column web thickness (after Morris).

Also, the stability of the outstand edge of the web stiffeners is ensured if:

Stiffener width/Stiffener thickness $= b_s/t_s \leqslant 10\sqrt{(275/p_y\,\text{stiffener})}$ (after Morris)

Longitudinal force due to abrupt stopping of the fully loaded crane could be carried by horizontal bracing from the underside of the gantry beam to a horizontal member between columns (at bays which are also braced vertically) so that yy moment is induced in this member which, of course, must be checked for combined stresses. Longitudinal force due to the wind may be resisted by bracing every sixth bay in the vertical and at rafter level. Gable columns may be provided at regular intervals considered pin ended (in Example 5.6 one column would suffice). The roof wind brace nodes should coincide with the gable column locations.

5.18 PORTAL FRAME BASES

The majority of portal framed buildings are designed to be pinned at the base; although a fixed base results in a saving of steelwork it results in greater expense for the foundations. However, fixed bases are often used for farm buildings and are the best solution where the side walls are designed to retain silage. Economy on pinned and fixed foundations can be achieved by placing a tie between the columns buried in the ground concrete slab (usually a round bar or small angle). For fixed bases the columns could be continued down past the tie (possibly 300 mm) with holding down bolts securing the column (with reduced moment at this elevation) to the pad foundation. In this case the tie position acts as a point of contraflexure. Calculation of the concrete base must include a check against base sliding (including active and passive soil pressure, and the soil coefficient of friction).

5.19 MULTI-BAY FRAMES

The design of a portal of several bays is generally based on the design of one frame. Plastic failure usually occurs in the end frame, but the design is only very slightly conservative if the single frame rafter sizes are used throughout. Internal columns may be designed as 'pinned' but the designer must *personally ensure* that a suitable note is added to the property deeds and that the owner is informed that removal of outer columns by accident or alteration would cause collapse of the entire structure. In areas subject to snow loading, a further separate case of snow loading should be considered to cater for drifting, based on the factors in BRE digest 290, or, if the project is outside the United Kingdom, based on ISO 4355: 1981.

5.20 SUMMARY

The plastic design method has been extended to include a wide variety of steel structures, such as grillages (flat plan support beams in a grid formation), vierendeel girders, and multi-storey frames.

The object of this chapter has been to present the reader with an appreciation of the basic ideas underlying plastic theory, with a typical structural frame design. However, for in-depth study and further examples the reader should refer to the classic literature on the subject [2, 5 and in particular 9].

5.21 PROBLEMS

5.1 A beam fixed at both ends has a span of 8m. It carries in load case 1 a uniformly distributed load of 6kN/m (factored) and in load case 2 a central point load of 20 kN (factored). Evaluate the maximum plastic moment.

5.2 A propped cantilever, span 7 m, carries a factored load of 7 kN/m. Calculate the plastic moment and select a suitable size of beam.

5.3 A continuous beam with uniform cross section has spans and loading (factored) shown on the diagram below. Evaluate the plastic moment for the critical span and select a suitable beam section.

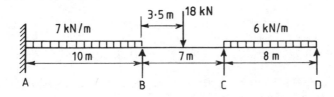

5.4 Determine the collapse moment for the rectangular portal frame shown overleaf. Factored loads are indicated.

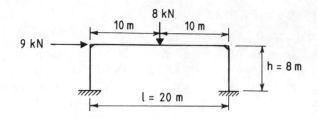

5.5 Determine the collapse moment of the pitched portal frame with fixed base as shown below. Purlin spacing at 1.25 m.

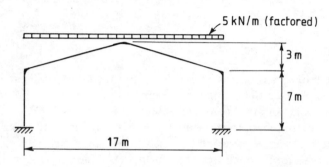

REFERENCES

1. Ghali, A. and Neville, A. M. (1978) *Structural Analysis.* Chapman and Hall, London.
2. Morris, L. J. and Randall, A. L. (1979) *Plastic Design.* Constrado, Croydon.
3. Goodwin, E. ( ) *Plastic Design of Steel Structures, Beams and Portals.*
4. Marshall (1978) *Solution of Problems in Structures.* Pitman, London.
5a. Baker, J. F. and Heyman, J. (1969) *Plastic Design of Frames, Vol. 1, Fundamentals.* Cambridge University Press.
5b. Heyman, J. *Plastic Design of Frames, Vol. 2, Applications.* Cambridge University Press.
6. Moy, S.S.I. (1981) *Plastic Methods for Steel and Concrete Structures.* Macmillan, London.
7. *Plastic Design in Steel* (1959) AISC.
8. *Commentary on Plastic Design in Steel* (1961) ASCE (American Society of Civil Engineers).
9. Baker, J. F., Horne, M. R. and Heyman, J. (1965) *The Steel Skeleton, Vol. II, Plastic Behaviour and Design.* Cambridge University Press.
10. Horne, M. R. and Morris, L. J. (1981). *Plastic Design of Low-Rise Frames.* Granada, London.
11. Morris, L. J. (1981) A commentary on portal frame design *Structural Eng.* December (and subsequent discussions).

Industrial buildings 6

6.1 INTRODUCTION

An industrial building is designed to accommodate the process which it is intended to house. The function of the building may range from warehouse storage to heavy machinery workshops with large electric overhead travelling

Fig. 6.1 Dinorwic pumped storage steelwork. The underground machine hall houses one of the world's most advanced hydroelectric projects and is the largest pumped storage power station in Europe. Client: C.E.G.B. Main contractor: MB2. Steelwork contractor: Robert Watson Ltd.

Fig. 6.2 Natwest Tower, London. Steelwork contractor: By kind permission of National Westminster Bank. Consulting engineer: Pell, Frischmann Consulting Engineers Ltd.

cranes. The basic aim is to produce a building that allows flexibility of layout and, if possible, a flexibility of use. If a future extension to the building is envisaged, one method adopted is to fabricate the end frames with connections designed to allow for future continuity.

At the preliminary design stage the possibility of unusual wind forces, earthquake, and possible ground subsidence, together with any special demands that the process itself places on the structure, such as impact loading or corrosive atmosphere, should be considered. Any one or combination of these effects may influence the final design adopted. When a framing system has been selected it is advisable to consider the possible effect of differential snow loading, i.e. snow building up on one side of a roof only.

6.2 FRAMING METHODS

Steel buildings are generally framed to resist forces and loads either by systems of bracing, or by a combination of bracing and framed moment connections. It is common to construct factory sheds with moment resisting portals at regular

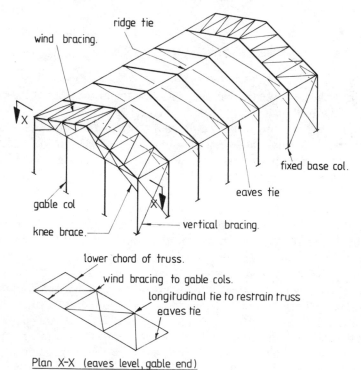

Fig. 6.3 General arrangement of a typical stanchion and truss frame.

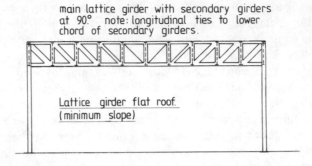

main lattice girder with secondary girders at 90° note: longitudinal ties to lower chord of secondary girders.

Lattice girder flat roof. (minimum slope)

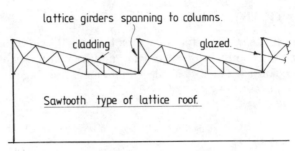

lattice girders spanning to columns.

cladding   glazed.

Sawtooth type of lattice roof.

Fig. 6.4 Lattice girder construction.

centres along the length of the building with bracing at the sides of the building and also in the roof plane.

The traditional method of providing roof cover is the stanchion and truss frame (Fig. 6.3). Although the steel weight in this type of design is often less than that for a comparable portal frame building the overall cost is generally higher because of the greater amount of labour involved in fabrication. Lattice construction is shown in Fig. 6.4.

The roof slope is often determined by drainage considerations. A minimum slope of about 1:100 is often used as a limit, introducing falls into the roof usually by means of sloping the top chord of lattice trusses. Large roof areas may be drained by positioning downpipes adjacent to internal columns. Occasionally the columns are constructed from tubes or RHS so that downpipes may be carried down inside the columns. The ideal slope for portal frame roofs probably lies between 19 and 30°, depending on the dimensions of the building and wind permeability (see CP3: Chapter V:Pt 2:1972). The roof slope can be varied in design to obtain minimum loading depending on the wind permeability of the building sides and roof. The majority of recently designed portal frames have a roof slope of from 6 to 12°, partly chosen because of the smaller volume of air involved in heating the building. Care must be taken with the cladding

Fig. 6.5 Portal frame construction. Warehouse, Lea Valley Industrial Estate. Consulting engineer: Pell, Frischmann Consulting Engineers Ltd.

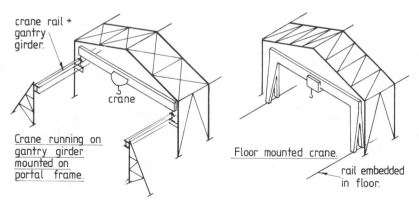

Fig. 6.6 Electric overhead travelling cranes in portal frame buildings.

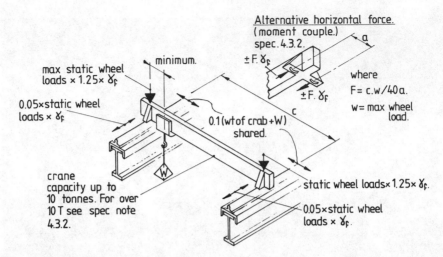

Fig. 6.7 Maximum crane loads for gantry girder and support structure.

details of 6° slope roofs, because of a tendency for rainwater to be blown up the roof slope. The steeper the roof slope of a portal, the greater the 'spread' at the knees, which is an undesirable distortion of the structure. When using overhead cranes, spread must be limited to a value specified by the crane manufacturer.

When overhead cranes are included in a frame there are several different ways of catering for the additional forces involved. For EOT cranes up to 10 tonnes (100 kN) the crane and transverse loads can generally be carried by a portal frame designed plastically (see Fig. 6.6). For larger cranes the frames should be designed elastically, and, if required, using wide column sections below the gantry girder, or alternatively latticed columns with 'fixed' bases. A crane running on floor rails may be the cheapest solution, unless the obstruction offered by the passage of the crane is considered to be a hazard or unacceptable due to workshop activity (Fig. 6.6). Maximum crane loads for a gantry girder are indicated in Fig. 6.7.

The spacing of frames in a building may depend on a variety of factors such as: clear spans required by client, maximum span for gantry girder (if applicable), or maximum span for purlins. Although it has become popular to place frames at spacings to match the maximum span of standard cold formed purlin shapes, lattice purlins may be found more economical for large spans. For a portal frame of 40 m span, 10 m spacing may be found economical; similarly, for a 60 m span, a 12 m spacing. A typical portal frame building is illustrated in Fig. 6.8.

Structures are sometimes required to carry vessels or equipment at various elevations, with conveyors or pipes running through the floors vertically and horizontally. The design of such structures demands a high exchange of

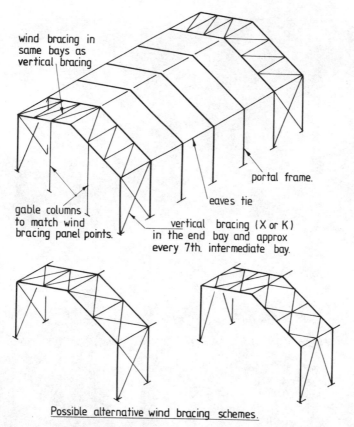

wind bracing in
same bays as
vertical bracing

portal frame.

gable columns
to match wind
bracing panel points.

eaves tie

vertical bracing (X or K)
in the end bay and approx
every 7th. intermediate bay.

Possible alternative wind bracing schemes.

Fig. 6.8 General arrangement of a typical portal frame building.

information between the process designers and the structural engineer, and the resulting structure is generally an evolved framework. Dynamic effects on the structure in this case may include hydraulic forces from the pipes, crane loads, and impact due to lifts or conveyors (see BS2655:Part 4 and BS5655: Part 1).

6.3 SPECIAL IMPOSED LOADING

In the design of chemical plant, the ordinary static and wind forces on the structure, vessels, and piping, are easily taken into account in the usual manner. However, additional forces usually occur on the pipes due to thermal expansion and hydraulic 'shock' loading. In the case of special structures carrying heaters or structures adjacent to heat sources, the effect of thermal expansion of the structure should be taken into account.

Thermal expansion (or contraction) of pipes will take place due to the

Fig. 6.9 Space frame construction. By kind permission of Watson Steelwork Fabricators.

prevailing solar conditions, or due to 'start up' temperature change when the plant is put into service. The horizontal 'friction force' caused by the lengthening (or shortening) of the pipes is often assumed to be about 10% of the weight of the pipe fluid in the service condition, but for instances where only one pipe rests on a support, 30% of the combined weight of pipe and fluid may be

Fig. 6.10 Clinker store, South Ferriby, for Rugby Portland Cement. Steelwork contractor: Robert Watson Ltd.

assumed. In addition to the pipe friction force, pipework is generally anchored at several positions along a piperack. In order to lessen the effect of pipe thermal expansion on the structure as a whole, large diameter pipes (i.e. pipes over about 300 mm dia.) are constructed in straight runs punctuated with horizontal U shapes. The longitudinal force on the pipe due to the expansion or contraction of the U shape is generally called the 'thermal force'. Anchorage points for both thermal and friction forces are generally positioned in the centre of straight runs of pipe, to balance forces as much as possible. The values of forces due to such pipework are generally supplied by piping or stress engineers.

Hydraulic 'shock' loading can occur due to unusual quantities of fluid travelling along the pipe and meeting a construction bend, or a vessel. Anchorage points for these pipes are positioned near to the expected point of shock load, with consideration given to reducing the force on welds at pipe connections.

Structures placed in hot or cold conditions will expand or contract with the change in ambient temperature. The steel selected must be suitable for the expected temperature range, possessing sufficient ductility for the lowest expected temperature to resist brittle fracture (see Chapter 1). The change in length of structural elements must be taken into account in the structural analysis, as an additional design case.

6.4 BRACING

Bracing members are usually intended to carry axial compression or tension. Planar bracing is commonly used, although the major exception is the bracing in space frames, which generally lies in several planes.

6.4.1 Cross and K bracing (illustrated as vertical bracing in Fig. 6.11)

These typical bracings cover the majority of variations. The method of design for K bracing is usually to halve the applied horizontal load F and design the brace for compression due to vertical load (if any) plus $F/2$. Cross bracing can be designed by assuming all the horizontal load F is carried by one member in tension. Thus smaller brace sizes can be achieved with cross bracing than with

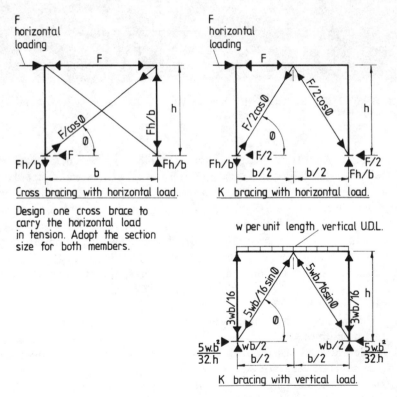

Cross bracing with horizontal load.

Design one cross brace to carry the horizontal load in tension. Adopt the section size for both members.

K bracing with horizontal load.

K bracing with vertical load.

Fig. 6.11 Typical cross and K bracings.

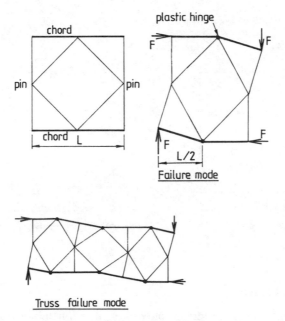

Fig. 6.12 Diamond bracing.

K bracing. Cross bracing members are generally connected at intersections, so reducing the effective length of the tension member. The disadvantages of cross bracing are that it creates greater obstruction in the braced area than K bracing, and horizontal load is carried by only one support, whereas K bracing carries half the load to each support. However, the addition of a tie/strut at the base of cross bracing ensures that the horizontal load is shared between supports.

6.4.2 Diamond bracing (see Fig. 6.12)

In situations where equipment or other items would interfere with more conventional bracing, diamond bracing may be used. Design of such arrangements must include a check that racking forces (if present) on the structure are not sufficient to cause the formation of plastic hinges in a manner similar to that shown in Fig. 6.12.

6.4.3 Knee bracing (illustrated as vertical bracing in Fig. 6.13)

Where headroom considerations preclude the use of K or cross bracing, knee bracing can be employed. Each brace is designed for the worst case compressive load. The use of this type of bracing can result in larger beam and column members because of the moments caused by the brace loads.

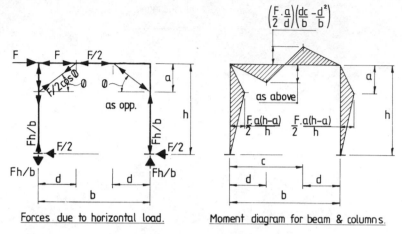

Forces due to horizontal load.   Moment diagram for beam & columns.

Fig. 6.13 Knee bracing.

6.4.4 Horizontal bracing

This type of bracing is generally employed in situations requiring resistance to wind forces. In the case of wind bracing in the roof of portal frame sheds, the bracing prevents longitudinal movement of, and eliminates horizontal bending in, the rafter.

In other types of building horizontal bracing may be designed in the form of lattice girders to resist horizontal forces without appreciable deflection. Structures adequately stiffened or braced in all faces could still be distorted on plan by eccentric forces, and bracing can also be designed for this case. Internal bracing for torsion on a rectangular tower is covered later in this chapter.

6.4.5. Lateral restraint bracing

In situations where steel members may be laterally unstable, it may be found necessary to use lateral restraint bracing (see Fig. 6.14). The most common

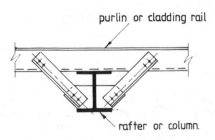

Fig. 6.14 Bracing to prevent lateral instability of a main member.

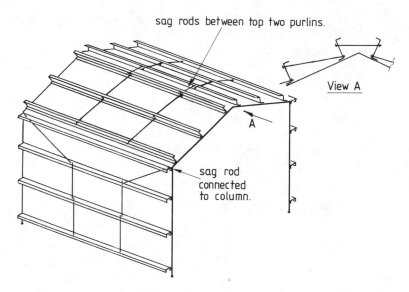

Fig. 6.15 The use of sag rods.

example of this bracing type is in portal frame buildings designed by the 'plastic' theory. At the portal eaves, a negative moment causes instability in the column and the rafter. Lateral restraint bracing is generally added at one or two cladding rail and purlin locations, designed for a force equal to 1% of the maximum factored force in the compression flange of the member, divided equally between the points of restraint.

6.4.6 Sag rods

Plain round rods are often used to support cladding rails or purlins at quarter or one third span points in order to limit bending stress or deflection in the weak axis. A typical arrangement is shown in Fig. 6.15. Heavier rods in

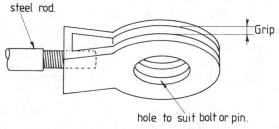

note : Grip = plate thickness + 6mm.

Fig. 6.16 Sketch of clevis.

suspension situations are connected with a clevis (Fig. 6.16), which is in turn bolted to a plate.

6.5 SPACE FRAMES

A 'space frame' is a three dimensional structure arranged in such a manner that the principal forces in the structure are axial, and bending in members is normally avoided. A space frame should carry all the imposed and dead loads safely to the foundations with the minimum amount of steelwork. Some typical space frames are:

- Towers (see Section 6.6)
- Shells, domes, arches, vaults
- Roof grids

Ideally, all space frames should be geodesic, having member node points on an 'invisible' solid of revolution, giving the most efficient structural form. However, at the present time it has generally been found more practical and economic to compromise between modular units and the space frame concept, using standard modules to build space frames. In the main, regular patterns or grids are used to 'frame' the space structure. Additional stiffeness is obtained when necessary by the use of double, or even triple, layers of a steel network. In the case of double layers the bottom chord lines of the lattice do not generally follow the top chord lines. Three-way grids (chords forming equilateral triangles on plan) represent the most economic form of framing for large spans, but generally require very accurate fabrication.

There are several different space frame systems currently available, each having particular characteristics and sometimes limitations. Four of these patented systems are:

(1) Spacedeck, (2) Mero, (3) Nodus, (4) Unibat

(1) *Spacedeck* is a British system, in use since 1950. It employs a square based (1.2 m) pyramid module with angles on the base and tube or bar bracing. In the system of construction the pyramids are inverted, connected together, and screwed adjustable members bolted to all the apices.
(2) *Mero* is a German system, in use since 1945. The system is based upon a prefabricated standard joint consisting of a sphere drilled and tapped for bracing bolts. The bracing consists of tubes, tapered each end, with movable studs which fit into the special joint. The angle of the holes on the sphere can be varied to suit any structural form.
(3) *Nodus* is a British system introduced by the British Steel Corporation Tubes Division, basically for the flat roof market. This system relies on a special joint consisting of steel castings with grooves to receive the special end connections welded to bracing. Hollow sections are used for main members.

(4) *Unibat* is a French system employing all structural section shapes, with no patented joint, or predetermined geometry. It is one of the few systems where large local bending can be catered for.

The selection of the most appropriate system depends upon the framing pattern adopted, the spans required, the type of loading, and the structural section shape. A careful comparison of cost should be carried out, but in general, structures with a high number of joints should be assembled in a system without special (expensive) joints, and for large span structures tubes should be used. Structures with local bending should use UB sections for the areas of bending.

The design of most space frames is carried out by computer. The calculations for bending moment (if applicable) and axial load are based on a series of simultaneous equations which are solved by means of matrices. The cost of structural analysis is related to the number of joints in the space frame. It is advisable to select a suitable structural grid with the minimum number of joints, symmetrical (about two axes) if possible so that only part of the structure need be analysed to determine the behaviour of the whole.

6.6 TOWERS

The type of tower selected will depend on several factors, such as use of tower, proposed height, location and ground elevation, material selection, deflection limitations and soil characteristics.

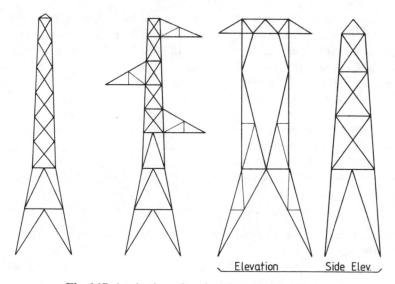

Fig. 6.17 A selection of typical free standing towers.

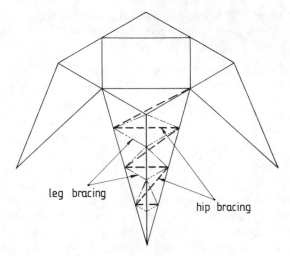

leg bracing

hip bracing

Fig. 6.18 Bracing the hip.

6.6.1 Free standing towers

A selection of typical free standing towers is illustrated in Fig. 6.17. Tall free standing towers are often tapered inwards on elevation and square at all sections on plan.

To avoid using large structural sections, lattice members are often employed. It is common to use lacing in towers to reduce the effective lengths of members in order that higher stresses may be used.

Significant torsional loads on a tower may be resisted by bracing adjacent faces to common points on the legs, otherwise additional bending in the legs would occur. The legs at the base of the tower are often stiffened by secondary bracing between a K brace and the leg; the K brace itself is stiffened by bracing the hip as shown in Fig. 6.18. K bracing is often used on the lower sections of towers, with X or single bracing used for the higher levels.

To find the forces in inclined legs, and in the bracing between the inclined legs due to horizontal loads on the tower, the legs are projected to cross at an imaginary node. Moments are then taken about the node and then required forces derived (Fig. 6.19). If the horizontal forces on the tower coincide with the imaginary node of the inclined legs then braces between the inclined legs would be zero loaded (apart from force due to restraint of the leg).

Torsion loads on a tower are divided into two components; i.e.

(1) A horizontal load on the centreline of the tower at the same elevation as the original load
(2) A horizontal moment on the tower at the same elevation as the original load.

The load is shared by the two parallel faces of the tower and the moment

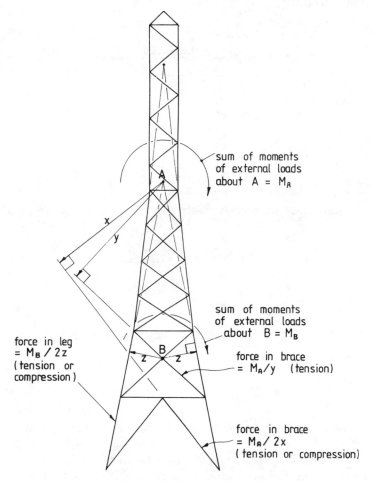

sum of moments
of external loads
about A = M_A

sum of moments
of external loads
about B = M_B

force in leg
= $M_B / 2z$
(tension or
compression)

force in brace
= M_A/y (tension)

force in brace
= $M_A / 2x$
(tension or compression)

Fig. 6.19 A typical tower design.

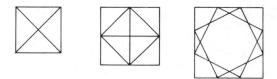

Fig. 6.20 Typical tower internal cross bracing.

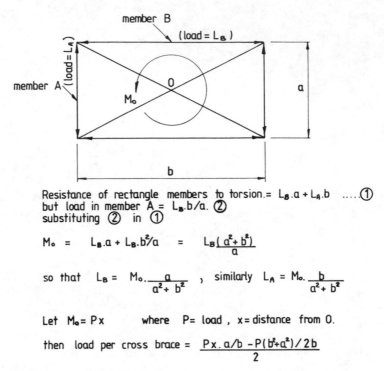

Resistance of rectangle members to torsion.$= L_B.a + L_A.b$ ....①
but load in member $A = L_B.b/a$. ②
substituting ② in ①

$$M_0 = L_B.a + L_B.b^2/a = \frac{L_B(a^2+b^2)}{a}$$

so that $L_B = M_0 . \dfrac{a}{a^2+b^2}$, similarly $L_A = M_0 . \dfrac{b}{a^2+b^2}$

Let $M_0 = Px$ where P= load , x= distance from 0.

then load per cross brace $= \dfrac{Px.a/b - P(b^2+a^2)/2b}{2}$

Fig. 6.21 Torsion on a rectangular tower.

shared by all faces. The rectangular form of the tower is maintained by internal cross bracing as shown in Fig. 6.20. Cross bracing is usually incorporated at cross arm levels and often at change of profile of tower. Torsion on a tower rectangular in plan is shown in Fig. 6.21. For a tower which is square on plan the torsion affects the bracing only, but for a rectangular tower the force in the leg can be ascertained by determining the force in each face and summing them.

Towers with a large number of redundancies such as those illustrated are best analysed by removing the redundancies and drawing a simple stress diagram. The following example has been calculated by the method of sections whereby approximate forces are obtained in order to size structural members, after which, for more exact analysis, a computer program may be used.

6.6.2 Analysis of a tower

Example 6.1

Analysis the forces due to wind in the tower shown in Fig. 6.22. Various members are selected to illustrate the design method employed.

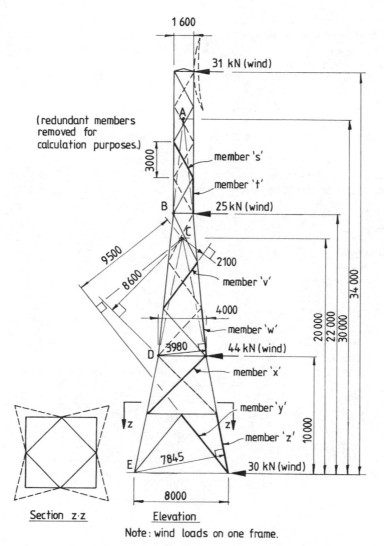

Fig. 6.22 Analysis of a tower – Example 6.1.

SOLUTION USING UNFACTORED LOADS

Member 's'

Angle of brace to leg $= \tan^{-1} (1.6/3.0) = 28.07°$

Force in member s $= 31 \sin 28.07° = 14.6$ kN tension or compression

Length of member $= \sqrt{(3^2 + 1.6^2)} = 3.4$ m

Member 't' – moments about B

Wind $1.6 F_t = (12 \times 31)$, $F_t = 232.5$ kN   T or C

Member 'v' – moments about C

Wind 2.1 $F_v = (2 \times 25) + (14 \times 31)$, $F_v = 230.4$ kN   T or C
Member 'w' – moments about D
Wind 3.98 $F_w = (12 \times 25) + (24 \times 31)$, $F_w = 262.3$ kN   T or C
Member 'x' – moments about C
Wind 8.6 $F_x = (31 \times 14) + (2 \times 25) - (10 \times 44)$, $F_x = 5.1$ kN   T
Member 'y' – moments about C
Wind 9.5 $F_y = (2 \times 25) + (31 \times 14) - (10 \times 44)$, $F_y = 4.6$ kN   C or T
Member 'z' – moments about E
Wind 7.845 $F_z = (34 \times 31) + (22 \times 25) + (10 \times 44)$, $F_z = 260.6$ kN   T or C

Note: Actual loads in the members will differ slightly from those shown above since dimensions used for lever arms have been taken from elevation and not from the diagrams showing the true (spatial) distances to members.

Example 6.2

Design of tower elements (Figs 6.23 and 6.24)
Design members 'x', 'y' and 'z' in the previous example.
Assume force due to dead load of tower, in member 'z' = 20 kN.

SOLUTION USING FACTORED LOADS
Member 'x' – tension member (2 bolts per connection or equivalent weld)
Loading = $1.4 \times 5.1 = 7.14$ kN tension   (see code Table 2)
Length of member = $\sqrt{(1^2 + 5^2 + 5^2)} = 7141$ mm

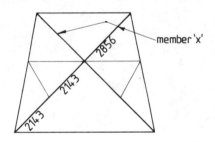

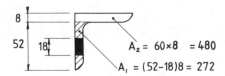

Fig. 6.23 Member 'x'.

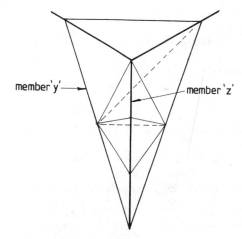

Fig. 6.24 Bracing the hip.

Try $60 \times 60 \times 8L$ and try arrangement indicated in Fig. 6.23.
Effective area $A_e = (A_1 + A_2)K_e$ (Fig. 6.23 and code 3.3.3)
where $A_1 + A_2 = $ net area, and $K_e = 1.2$ for grade 43 steel.
Max. allowable load in tension $= p_y A_e$ (code 4.6.1)
$$= 0.275(272 + 480)1.2$$
$$= 248\,kN > 7.14\,kN$$
Use $60 \times 60 \times 8$ in arrangement as shown in Fig. 6.23

Member 'y' – tension or compression member – design as compression member.
Loading $= 1.4 \times 4.6 = 6.5\,kN$
Length of member $= \sqrt{(1^2 + 4^2 + 5^2)} = 6.4\,m$; try $60 \times 60 \times 8L$ bracing the hip as shown in Fig. 6.24, to reduce effective length. Then
$\lambda = L_e/r_{vv} = 3200 \times 0.85/11.6 = 234$
and $p_c = 31.0\,N/mm^2$ (code Table 27(c))
Max. allowable load on member $= 31 \times 9.03 \times 10^2/10^3 = 28\,kN > 6.5$
Use $60 \times 60 \times 8L$ in arrangement as shown in Fig. 6.24

Member 'z' – tension or compression member – design as a strut
Dead load at member 'z' $= 20\,kN$
Worst loading case is wind $+$ dead $= 1.4(260.6 + 20) = 393\,kN$
Length of member $= \sqrt{(1^2 + 1^2 + 5^2)} = 5196\,mm$
Effective length $= 1732 \times 0.85 = 1472\,mm$
Try $120 \times 120 \times 10L$
$\lambda = L_e/r_{vv} = 1472/23.6 = 62$
$p_c = 197\,N/mm^2$ (code Table 27(c))
Max. allowable load on member $= 197 \times 23.2 \times 10^2/10^3 = 457\,kN > 393$
Use $120 \times 120 \times 10\,L$

REFERENCES

1. Bamber, D. (1957) *The Design of Towers, Masts and Pylons.* AESD Pamphlet.
2. Stainsby, R. (1980) Automation in the fabrication and design of welded beams, girders and columns. Structural Eng. **58A**, May.
3. Clift, N. and Meeke, I. B. (1980) Developments in the design of buoyant ferry ramps. *Structural Eng.* **58A**, March.
4. Dowling, P. J., Mears, T. F., Owens, G. W. & Raven, G. K. (1982) A development in the automated design and fabrication of portal framed industrial buildings. *Structural Eng.* **60A**, October.
5. Clague, K. and Wright, H. (1973) Pressure in bunkers. *Iron and Steel International* August.
6. Blodgett, O. W. (1980) Detailing to achieve practical welded fabrication *AISC*, 4th Quarter.

7.1 INTRODUCTION

The erection of a steel structure should proceed in a planned sequence. Bar charts are often used as a means of recording erection progress. In addition to the set of drawings required for erection of the structure, a site bolt list should be provided which generally includes an allowance for possible loss of bolts.

Delivery of structural steel should be scheduled to coincide with the erection sequence. Timber baulks are usually placed at suitable centres in the site storage area to ensure that the stored steel is clear of the ground. Additional baulks are used to separate steel members to avoid damage to cleats. If possible the steel should be stored in a site location that avoids double handling.

The resident engineer and the contractor should familiarize themselves with the safety guidelines given in CP3010, and BS5531.

7.2 DELIVERY OF STEELWORK

Delivered steelwork should be inspected for completeness in accordance with the delivery schedule and the total weight verified by using a weighbridge. All members should be free from accidental damage. It is advisable to ensure that loose fittings such as packs are securely wired to the members. Damage to paintwork is usually repaired before the steelwork is erected. The steelwork members should be clearly marked by the fabricator in accordance with a marking plan. This enables the erection contractor to place each member in its correct location on the structure. The list of delivered mark numbers should be retained with the delivery note as a record of delivery, to assist in the storage and selection of the parts for erection.

7.3 SETTING OUT AND ERECTION OF STEELWORK

On large sites a number of surveying 'stations' of known location and elevation are used as local reference points, and on all sites levels are checked in relation to the ordnance survey datum levels. The locations of bases are usually set-out and checked by the use of a theodolite. The height of the bases should be constructed accurately, and the columns are levelled by the insertion of steel shim plates and grout between the top of foundation and column base plate.

It is preferable to allow the foundation bolts to project more than the basic minimum to assist in counteracting inaccurate levelling of the bases. The column

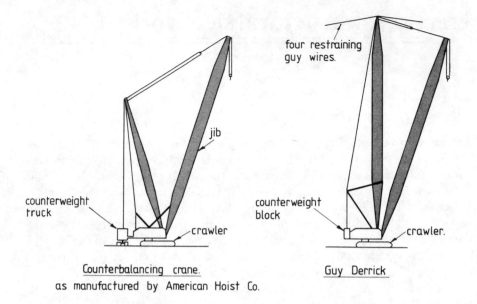

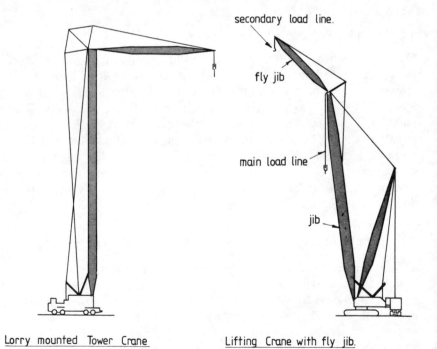

Fig. 7.1 Types of crane.

Table 7.1. Permissible normal tolerances.

ECCS recommendations (extracted from the European Recommendations for Steel Construction)			BS 5950: Part 2
Deviation		a_{max}	
Overall dimensions of the building	$\Sigma \Delta h$ $\Sigma \Delta l$	± 20 mm for $l \leqslant 30$ m $\pm 20 + 0.25\,(l-30)$ mm for 30 m $< l \leqslant 210$ m	± 10 mm for plan location of column.
Level of top of floor slab Floor bearing on column	Δh	± 5 mm	± 5 m within local radius of 5 metres, ± 10 mm for entire floor.
Inclination of column (a) between adjacent floor slabs (height h_1)	v_h	$0.0035\,h_1$	$0.0017\,h_1$ but 5 mm max.
(b) maximum deviation for the vertical line through the column base	v_1	$0.0035\,(\Sigma h_1)\,3/(n+2)$	5 mm per storey.
Deflection of column between floor slabs (height h_1)	f_h	$0.0015\,h_1$	—
Lateral deflection of girder (span l_b)	f_1	$0.0015\,l_b$ $\leqslant 40$ mm	—
Unintentional eccentricity of girder bearing	e_0	5 mm	5 mm
Distance between adjacent steel columns at every section	Δl_s	± 15 mm	—
Distance between adjacent steel girders at every section	Δl_t	± 20 mm	
Welded girders and columns (depth of web h_w, width of flange b):			—
Deflection of web	f_w	$h_w/150$	
Inclination of web between upper and lower flange	v_w	$h_w/75$	
Eccentricity of the web in relation to the centre of either flange	v_{wl}	$b/40$ $\leqslant 10$ mm	

centrelines are marked on the steel baseplate generally with a black line on white paint background.

It is the usual procedure to erect the steelwork from a fully braced 'box' which ensures greater safety against accidental collapse, and an accurate datum from which to ensure plumbness and level. Additional bracing is often needed during erection to obviate torsional instability, and to ensure the structural stability of the partly completed structure.

7.4 CRANAGE

The engineer normally inspects and approves the proposed method of erection to be carried out by the contractor. Where cranes are to be used, it is normal for the contractor to provide a plan showing the intended crane location and boom reach. Steel structures such as portal frames are erected by means of mobile hydraulic cranes, but larger structures such as power stations are generally erected by means of tall mobile cranes or tower cranes (Fig. 7.1).

The engineer should check that the cranes have test certificates, and that they are free from structural damage.

7.5 CONSTRUCTION TOLERANCES

The accuracy of construction depends on mill rolling tolerances, fabrication tolerances, and the tolerances of erection. An acceptable limit of deviation from straightness for main members is 1/1000 of the length between points which are to be laterally restrained, columns faced at each end for bearing must have a tolerance on length of ± 1 mm. Beams and other members should have a tolerance on length of $+0$ and -4 mm. The negative tolerance on beams enables the erector to correct errors by the addition of small packs. Precast concrete and plastic fascia panels should be manufactured with slotted connections to allow for cumulative tolerances. An inspection should be made of the mill rolling tolerances on structural shapes, such as column sections out of shape. An out of shape column section may impede the erection of precast or fascia panels.

Temperature variations during erection must be taken into account by the steelwork contractor. Steel expands or contracts at the rate of ± 3.6 mm per 30 m for every $\pm 10°C$ change. Large structures (such as tall buildings) should be erected using a central setting out point based on a true plumb to a column base work point. Assuming an ambient temperature datum of, say, 15°C in a temperate climate, adjustments to horizontal dimensions are made at the time of erection.

The vertical plumbness of the whole of a column length is limited to one thousandth of the length or 5 mm per storey, whichever is the greater, although in the case of very high multi-storey buildings greater accuracy is required.

Fig. 7.2 Washery Building – Prince of Wales Colliery. Steelwork contractor: Robert Watson and Co. (Constructional Engineers) Ltd. This photograph illustrates the use of a derrick crane and a crawler crane.

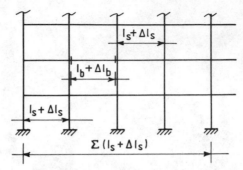

Fig. 7.3 Deviation in length (vertical section).

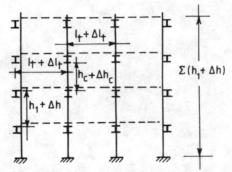

Fig. 7.4 Deviations in height and length.

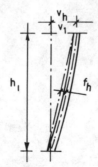

Fig. 7.5 Inclination and deflection.

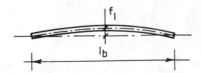

Fig. 7.6 Lateral deflection of girders.

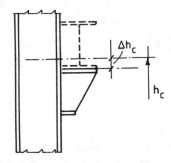

Fig. 7.7 Eccentricity of girder bearing.

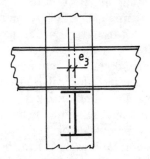

Fig. 7.8 Deviations in welded girders.

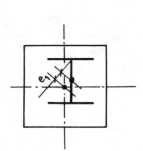

Fig. 7.9 Deviation in connecting pieces.

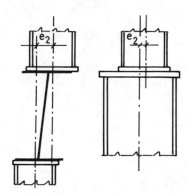

Fig. 7.10 Deviations in column splices.

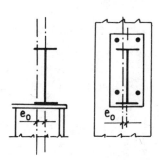

Fig. 7.11 Deviation in level of bearing surfaces.

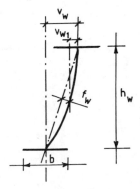

Fig. 7.12 Deviation in position of bearing surfaces.

Fabrication and erection tolerances are prescribed in BS5950: Part 2 and they are compared with ECCS recommendations in Table 7.1. Figures 7.3 to 7.12 are reproduced from the European recommendations.

Distortion of the structure may occur due to the mode of erection. Differential shortening of internal columns can occur on a very tall building during erection due to the fact that internal columns carry higher axial load than external columns. This shortening could cause connecting beams to be fixed in position out of level. Ideally, on such structures, beams should be fixed with a slope towards the outside of the building, so that when final finishes are placed, the steel beams will level, due to the extra dead loading. When cover to the steel beams is small such adjustments become a major consideration. Precast facades should be progressively placed on an entire floor before proceeding with the next floor. This procedure should minimize the possibility of the steel frame being pulled out of alignment.

7.6 FINISHING OF ERECTED STEELWORK

The lined and levelled steelwork must be inspected for proper bolting (especially important in the case of HSFG bolts) and damage to paintwork. The retouching of paint should be completed with the same paint system as used previously. Damage to galvanized items is difficult to remedy but repair can be effected by grit blasting and zinc spraying. All bolts should be painted or otherwise protected by a metallic coating.

REFERENCES

1. Merritt, F. S. (1983) *Standard Handbook for Civil Engineers*. McGraw-Hill, USA.
2. AISC (1980) *Manual of Steel Construction*, 8th. edn. American Institute of Steel Construction, New York.
3. (1981) Site inspection of structural steelwork. *Proc. Instn. Civ. Engrs.* Part 1, **70**, pp. 390–402.
4. BS5531 (1978) *Code of Practice for Safety in Erecting Structural Frames*. BSI, London.
5. CP3010 (1972) *Safe Use of Cranes (Mobile Cranes, Tower Cranes, and Derrick Cranes)*. BSI, London.
6. ECCS (1984–1986) *European Recommendations for Steel Construction*. European Convention for Constructional Steelwork.
7. Various authors (1983) *Rigging for Commercial Construction*. Reston Publishing Inc. Virginia.

Multi-storey building – Design example   **8**

8.1 INTRODUCTION

There are three concepts used when designing multi-storey steel structures.

(1) Simple design
(2) Semi-rigid design
(3) Fully rigid design.

The simple design concept assumes that the connections between members will not develop restraint moments and are therefore assumed to be pin jointed.

Fig. 8.1 Building for Marks and Spencer, Caernarfon. Steelwork Contractor: Modern Engineering (Bristol) Ltd.

For simple design the following assumptions are made:

- The beams are simply supported
- All connections of beams to columns are flexible and are proportioned for the reaction shears, applied at the appropriate eccentricity.

In order to accommodate horizontal forces (e.g. wind), side sway can be prevented by inserting bracing, or by utilizing concrete shear walls, lift or stair enclosures, acting together with the shear resistance of the floor slabs. The bracing and shear wall concepts are both often used.

A semi-rigid design permits a reduction in the maximum bending moment in beams suitably connected to their supports. The connections are designed to provide some direction fixity. This method is not recommended by the authors, due to a lack of moment/rotation data of standard type beam/column connections.

In fully rigid design the connections between members must have sufficient strength and rigidity to justify the assumption that the original angles between the connected members remain virtually unchanged when loads are applied. When a plastic analysis is utilized the strength of the connection is vital and should be at least equal in strength and rigidity to the connected members. These connections are generally referred to as 'moment' connections.

For fully rigid design or continuous construction the loads on members should

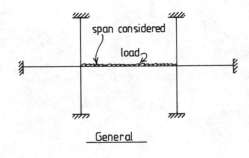

General

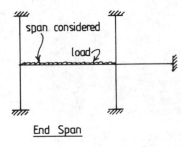

End Span

Fig. 8.2

be arranged in the most unfavourable pattern in order to obtain the maximum/minimum forces and moments.

For a multi-storey frame the number of load cases are numerous. Therefore the code recommends the use of sub-frames to facilitate analysis. Examples of sub-frames are shown in Fig. 8.2.

8.2 REDUCTION OF IMPOSED FLOOR LOAD

Imposed or non-permanent loads for different building occupations are obtained from BS6399: Part 1: 1984.

For multi-storey buildings it is most unlikely that all floors will be loaded at the same time. Therefore BS6399 allows a reduction in total distributed imposed loading. The allowable reductions are shown in Table 8.1.

An alternative reduction of imposed loading allowed is that where a single span of beam supports not less than $40\,m^2$ of floor at one level, the imposed load for beam design may be reduced by 5% for each $40\,m^2$ supported, subject to a maximum reduction of 25%. This reduction, or that stated in Table 8.1, whichever is greater, may be utilized for the design of beams, columns, walls and foundations.

The above-mentioned reductions of imposed loading should not be used for buildings such as warehouses, garages and offices or other areas which are used for storage and filing purposes.

The design of a five storey office building in structural steel using the simple design concept follows.

Generally the steel frame (floor beams and columns) has to be protected against fire. The method of protection depends upon the fire rating required by the building regulations, cost of protection and ease of erection.

Table 8.1 Reduction in total distributed imposed floor loads

Number of floors, including the roof, carried by member under consideration	Reduction in total distributed imposed load on all floors carried by the member under consideration
	%
1	0
2	10
3	20
4	30
5 to 10	40
over 10	50

Note: Where the floor is designed for $5\,kN/m^2$ or more, the above reductions may be taken provided that the loading assumed is not less than it would have been if all floors had been designed for $5\,kN/m^2$ with no reductions.

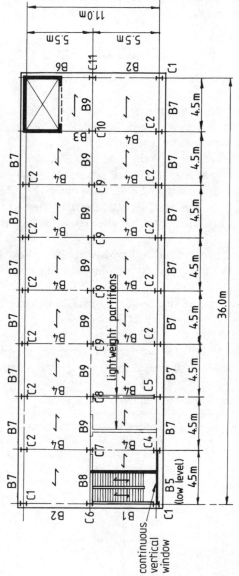

Fig. 8.3 First to fourth floor plan.

There are various methods of fire protection which include:

(1) Concrete casing (*in situ* or precast)
(2) Lightweight casing, forming a box around each member
(3) Suspended ceilings
(4) Applied coatings often sprayed around the profile of the member.

In the following example both uncased and cased members have been analysed to illustrate the different design methods.

8.3 OFFICE BUILDING

A typical plan and sections of the building are shown in Figs 8.3 to 8.8. Construction details and finishes are listed below.

Roof	Asphalt, lightweight screed, concrete slab, suspended ceiling.
Typical floor	Tiles, screed, concrete slab, suspended ceiling.
Ground floor	Screed, concrete slab on hardcore.
Steel frame	Roof beams Lightweight casing
	Floor beams Cased in concrete } For 1 hour
	Columns Cased in concrete } fire protection
External walls	Precast concrete cladding with continuous windows over.
Internal walls	Concrete walls around lift shaft and staircase well. (These walls

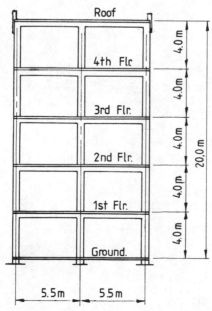

Fig. 8.4 Typical cross section.

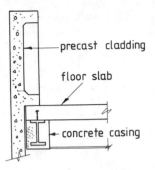

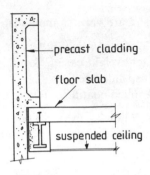

Section through end elevation.
1st. to 4th. floors.

Section through longitudinal
elevation.
1st. to 4th. floors.

Fig. 8.5 Typical details of office building.

Fig. 8.6

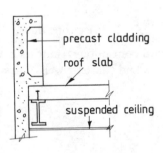

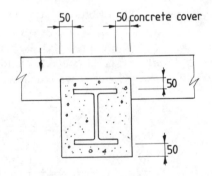

Section through end elevation
Roof.

Plan on external column.

Fig. 8.7

Fig. 8.8

are designed as shear walls to resist the wind on the building.)
Lightweight partitions to subdivide offices where required.

Imposed floor loading As stated in BS6399: Part 1.

Wind loading As stated in CP3: Chapter V: Part 2.

8.4 ROOF BEAMS

8.4.1 Roof loading

Dead load (from BS648) kN/m^2
  Asphalt 0.5

Lightweight screed	0.5
Concrete slab	2.6
Steel beam	0.4
Suspended ceiling	0.4
	———
Total dead load	4.4
Imposed load	2.5
	———
Total loading	$6.9 \, kN/m^2$

External cladding $1.6 \times 0.1 \times 2.4 = 0.4 \, kN/m$
Steel grade 43.

8.4.2 Roof beam design

Beam	Span m	Calculation	R_L kN	R_R kN	S_x cm³	Selected section	Code ref.
B2	5.50	D.L. ($\gamma_f = 1.4$)					Load factors from Table 2
B1 similar		Roof slab, $5.5 \times 2.25 \times 4.4 \times 1.4$ etc. $= 76.2$					
		Wall unit $5.5 \times 0.4 \times 1.4$ $= \dfrac{3.1}{79.3\ \text{kN.}}$					
		L.L. ($\gamma_f = 1.6$)					
		$5.5 \times 2.25 \times 2.5 \times 1.6 = 49.5$					
		Total $= 128.8$	64.4	64.4			
		Beam laterally restrained by conc. slab.					
		Use max. allowable $p_b = 275\ \text{N/mm}^2$					
		$M = WL/8 = 128.8 \times 5.5/8 = 88.6\ \text{kNm}$					
		Min. $S_x = M/p_b = 88.6 \times 10^3/275$			395.6	$254 \times 146 \times 31$ UB compact section	Table 7
		$= 322 < 395.6$					
		Check deflection					Table 2.1 Chapter 2 (Deflection restricted to $L/360$)
		$M_{\text{L.L.}}$ (unfactored)					
		$30.9 \times 5.5/8 = 21.2\ \text{kNm}$					
		Min. $I = 18.2\ MI(\text{UDL})$					
		$= 18.2 \times 21.2 \times 5.5 = 2122\ \text{cm}^4$					
		(provided $I = 4427\ \text{cm}^4$)					
		Shear Max. $F_v = 128.8/2 = 64.4\ \text{kN}$					
		Shear capacity (P_v)					4.2.5.
		$A_v = tD = 6.1 \times 251.5 = 1534\ \text{mm}^2$					
		$P_v = 0.6 \times 275 \times 1534/10^3 = 253\ \text{kN}$					
		$F_v \leqslant 0.6\,P_v \leqslant 152\ \text{kN}$					
		$M_c = p_y S_x = 275 \times 395.6/10^3$					
		$= 109 > 88.6\ \text{kNm}$					

4.2.2.

B4

For full lateral restraint, slab should be capable of resisting a lateral force of 1% of max. force in compression flange of beam using factored loading.

5.50

$$\begin{array}{ll} & \text{kN}\\ \text{D.L. } 5.5 \times 4.5 \times 4.4 \times 1.4 = & 152.4\\ \text{L.L } 5.5 \times 4.5 \times 2.5 \times 1.6 = & \underline{99.0}\\ \qquad\qquad \text{Total} = & 251.4 \end{array}$$

125.7   125.7

Laterally restrained by slab

$M = WL/8 = 251.4 \times 5.5/8 = 172.8\,\text{kNm}$

Min. $S_x = 172.8 \times 10^3/275 = 628\,\text{cm}^3$

Check deflection

$M_{L.L}$ (unfactored) $= 61.9 \times 5.5/8 = 42.6\,\text{kNm}$

Min. $I = 18.2\,ML$

$= 18.2 \times 42.6 \times 5.5 = 4264\,\text{cm}^4$

(provided $12091\,\text{cm}^4$)

773.7 356×171 UB 45 (compact)   Table 7

Shear force at support

$F_v = 125.7\,\text{kN}$

$A_v = tD = 352 \times 6.9 = 2429\,\text{mm}^2$

$P_v = 0.6 \times 275 \times 2429/10^3 = 400\,\text{kN}$

$F_v \leqslant 0.6 \times 400 = 240\,\text{kN}$

$M_c = p_y S_x = 275 \times 773.7/10^3 = 213\,\text{kNm}$

B9

4.50

Beams span in same direction as slab. They should be checked to carry a nominal width of slab (say 1m).

0.8   0.8 152×89 Joist $\times$ 17.09

Note: These beams act as ties; they and their connections should be capable of carrying a factored tensile load of 40 kN. (Ties at floor levels – 75 kN.)

2.4.5.2

Roof beam design (continued)

Beam	Span m	Calculation	R_L kN	R_R kN	S_x cm³	Selected section	Code ref.
B7	4.50	D.L. kN wall unit $= 0.4 \times 4.5 \times 1.4$ $= 2.5$					
Edge beam		slab D.L. $= 1 \times 5.5 \times 4.4 \times 1.4$ $= 33.9$ (nominal) L.L. $1.1 \times 5.5 \times 2.5 \times 1.6 = 24.1$ Total $= 60.5$	30.3	30.3			
		$M = 60.5 \times 4.5/8 = 34$ kNm beam compression flange restrained use $p_b = 275$ N/mm²					
		check if section is compact $b/T = 50.8/9 = 5.6 < 9.5\sqrt{(275/275)}$ $\therefore$ Section compact Min. $S_x = M/p_b = 34 \times 10^6/275 \times 10^3 = 124\,\text{cm}^3$ $< 193\,\text{cm}^3$ provided			193	$178 \times 102 \times 21.5$ Joist	Table 7

Notes. These beams form a continuous horizontal peripheral tie at each floor level and the code (2.4.5.3(b)) requires that:

(1) For multi-storey buildings designed to localize accidental damage, ties should be arranged in continuous lines throughout each floor and roof and should be checked for the following factored tensile loads (not considered additive to other loads):

     Internal ties $0.5\,w_f s_t L_a$ $\Bigg\}$ but not less than 40 kN for roof, and

     External ties $0.25 w_f s_t L_a$ not less than 75 kN for floors

where $w_f =$ total factored dead and imposed load per unit area

     $s_t =$ mean transverse spacing of the ties

     $L_a =$ the greatest distance, in direction of the tie, between vertical supports.

(2) Ties anchoring columns at periphery of floor or roof should be checked for the force obtained in (1) above, but shall not be less than 1% of the factored vertical load on the column at the level considered.

230

8.5 1ST TO 4TH FLOOR BEAMS (CASED)

8.5.1 Floor loading

Dead load	kN/m²
Tiles	0.1
Screed	0.9
150 Concrete slab	3.6
Steel beams	0.5
Suspended ceiling	0.4
Partitions (lightweight)	1.0
Total dead load	6.5
Imposed load	5.0
Total loading	11.5 kN/m²

External cladding 0.4 kN/m

8.5.2 1st to 4th floor beams – designs

Beam	Span m	Calculation	R_L kN	R_R kN	S_x cm³	Selected section	Code ref.
B2	5.5	D.L. Floor $6.5 \times 5.5 \times 4.5/2 \times 1.4$ $\quad$ = 112.6 (kN) p.c. cladding $0.4 \times 5.5 \times 1.4$ = 3.1 $\quad$ 115.7 L.L. Floor $5 \times 5.5 \times 4.5/2 \times 1.6$ = 99.0 $\qquad$ Total load = 214.7 $M = 214.7 \times 5.5/8 = 148$ kNm Beam restrained by floor slab; use max. allowable $p_b = 275$ N/mm² Min. $S_x = 148 \times 10^6/275 \times 10^3 = 538$ cm³ S_x provided 653.6 cm³ Shear Max. $F_v = 214.7/2 = 107.4$ kN $A_v = tD = 6.5 \times 352.8 = 2293$ mm² Allowable shear capacity $P_v = 0.6 \times 275 \times 2293/10^3 = 378$ kN $F_v \leqslant 0.6 \times 378 = 227$ kN and $M_c = p_y S_x \leqslant 1.2 p_y Z$ $\quad = 275 \times 653.6/10^3 = 180$ kNm	107.4	107.4	653.6	356 × 127 × 39 UB Compact section	Table 7
B1	5.5	Check deflection applying unfactored live load. For procedure refer to roof beam B2. Staircase spans between beams B5 and B8; beam B1 supports cladding only and is not restrained by floor slab. The beam is concrete cased and is designed as a cased beam.					

D.L. (cladding and casing)

$(0.4 + 2.5)5.5 \times 1.4 = 22.3$ kN   11.2   11.2

$M = 22.3 \times 5.5/8 = 15.3$ kNm

Min. $S_x = 15.3 \times 10^3/275 = 55.6$ cm^3 539.8 $356 \times 127 \times 33$ UB   4.14.1(i)

The effective length of a cased beam $\ngtr$ the smallest of

(1) $40b_c = 40 \times 225 = 9000$ mm

(2) $100b_c^2/d_c = 100 \times 225^2/410 = 12347$ mm

(3) $250r = 250 \times 25.9 = 6475$ mm

where r = minimum radius of gyration of steel beam only and b_c, d_c are as shown in sketch

6475 mm $>$ actual eff. length so use

$L_e = 5.5$ m.

Radius of gyration of cased beam   4.14.2

$= 0.2(B + 100) = 0.2(125 + 100)$

$= 45$ mm > 25.9

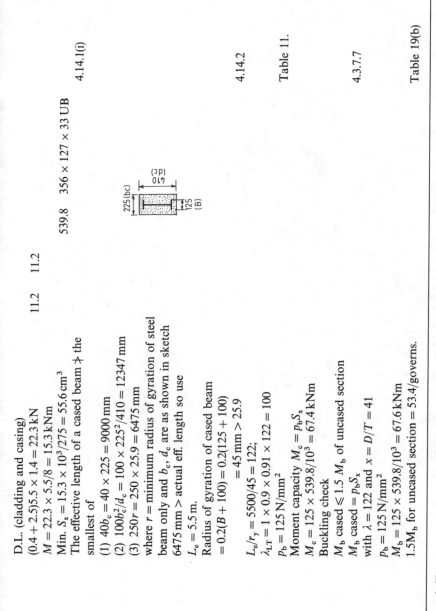

$L_e/r_y = 5500/45 = 122$;

$\lambda_{LT} = 1 \times 0.9 \times 0.91 \times 122 = 100$ Table 11.

$p_b = 125$ N/mm^2

Moment capacity $M_c = p_b S_x$

$M_c = 125 \times 539.8/10^3 = 67.4$ kNm

Buckling check

M_b cased $\leqslant 1.5\ M_b$ of uncased section

M_b cased $= p_b S_x$ 4.3.7.7

with $\lambda = 122$ and $x = D/T = 41$

$p_b = 125$ N/mm^2

$M_b = 125 \times 539.8/10^3 = 67.6$ kNm   Table 19(b)

$1.5M_b$ for uncased section $= 53.4$/governs.

Note: This example illustrates the design of a concrete cased beam. The beam size selected is of adequate strength without the casing (casing required for fireproofing). Beams to be solidly encased in dense concrete with a characteristic cube strength at 28 days of not less than 20 N/mm^2 (grade 20). Concrete casing to be reinforced using steel wire fabric to BS4483 ref. D98 or steel bars not less than 5 mm diameter may be used in form of closed links at not more than 200 mm centres attached to 4 longitudinal spacing bars. The reinforcement to be placed in centre of concrete cover. Concrete cover to be not less than 50 mm from outer face and edges of steel member.

Floor beams (continued)

Beam	Span m	Calculation	R_L kN	R_R kN	S_x cm³	Selected section	Code ref.
B4	5.50	D.L. Casing $3.4 \times 5.5 \times 1.4$ $= 26.2$ Floor slab $6.5 \times 5.5 \times 4.5 \times 1.4$ $= 225.2$ L.L. $5.0 \times 5.5 \times 4.5 \times 1.6$ $= 198.0$ Total $\underline{449.4}$					
			224.7	224.7			
		$M = 449.4 \times 5.5/8 = 309$ kNm $p_b = 275$ N/mm² $S_x = 309 \times 10^6/275 \times 10^3 = 1124$ cm³ Top flange of beam restrained by slab. Check for compact section $b/T = 76.5/13.3 = 5.75 < 9.5 \sqrt{(275/275)}$ ∴ compact section Shear $F_v = 224.7$ kN $A_v = 8 \times 454.7 = 3637$ mm² $P_v = 0.6 \times 275/10^3 \times 3637 = 600$ kN $F_v < 0.6 \times 600 = 360$ kN $M_c = 275 \times 1284/10^3 = 352$ kNm > 309 Check deflection using unfactored loading.			1284	457 × 152 × 60 UB	Table 7
B9	4.50	These beams, parallel to span of slab, act as ties. Check beam section by applying a nominal width of slab load. Each tie shall be able to resist a factored tensile force of 75 kN (at each floor level). For multi-storey buildings designed to localize accidental damage refer to notes after roof beam B7	5.0	5.0		254 × 114 × 37.2 Joist	

I-section diagram: width 555, depth 255.

Hall staircase loads are carried

Reaction of stair load (unfactored)
= 18.38 kN/m (over 2.5 m)

                             kN/m

D.L. Stairs 18.38×1.4 = 25.7
L.L. Stairs $5 \times 4.5/2 \times 1.6$ = 18.0
                                  43.7

D.L. Beam + casing 2.5×1.4 = 3.5

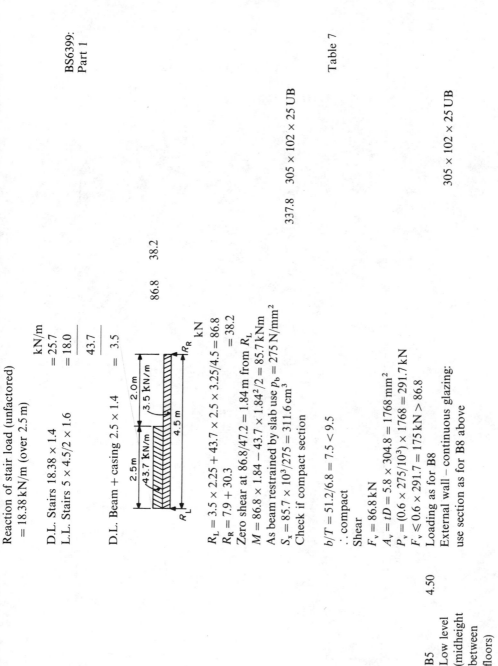

$R_L = 3.5 \times 2.25 + 43.7 \times 2.5 \times 3.25/4.5 = 86.8$
$R_R = 7.9 + 30.3$ $= 38.2$
Zero shear at $86.8/47.2 = 1.84$ m from R_L
$M = 86.8 \times 1.84 - 43.7 \times 1.84^2/2 = 85.7$ kNm
As beam restrained by slab use $p_b = 275$ N/mm²
$S_x = 85.7 \times 10^3/275 = 311.6$ cm³
Check if compact section

     86.8   38.2

337.8   305 × 102 × 25 UB

BS6399:
Part 1

Table 7

$b/T = 51.2/6.8 = 7.5 < 9.5$
∴ compact
Shear
$F_v = 86.8$ kN
$A_v = tD = 5.8 \times 304.8 = 1768$ mm²
$P_v = (0.6 \times 275/10^3) \times 1768 = 291.7$ kN
$F_v \leqslant 0.6 \times 291.7 = 175$ kN > 86.8

B5   4.50   Loading as for B8
Low level   External wall – continuous glazing:
(midheight   use section as for B8 above
between
floors)

305 × 102 × 25 UB

235

Floor beams (continued)

Beam	Span m	Calculation	R_L kN	R_R kN	S_x cm³	Selected section	Code ref.
B7 Edge beam	4.50	D.L. kN p.c. wall unit $0.4 \times 4.5 \times 1.4 = 2.6$ 1 metre width (nominal) slab D.L. $1 \times 4.5 \times 6.5 \times 1.4 = 41.0$ L.L. $1 \times 4.5 \times 5.0 \times 1.6 = 36.0$ Total <u>79.6</u> Beam compression flange restrained Use $p_b = 275\,\text{N/mm}^2$ $M = 79.6 \times 4.5/8 = 45\,\text{kNm}$ Min. $S_x = 45 \times 10^3/275 = 164\,\text{cm}^3$ Shear $F_v = 40\,\text{kN}$ $A_v = tD = 5.8 \times 203.2 = 1178\,\text{mm}^2$ $P_v = (0.6 \times 275/10^3) \times 1178 = 194\,\text{kN}$ $F_v \leqslant 0.6 \times 194 = 116\,\text{kN} > 39.8$	39.8	39.8	259.8	$203 \times 133 \times 25$ UB (compact)	

8.6 COLUMNS (CASED)

8.6.1 Notes for cased columns

Cased struts

Radius of gyration (r_y) about its axis in plane of web taken
as $0.02b_c$ but not more than $0.2\,(B + 150)$ mm   (code 4.14.3(a))
where b_c = min. width of solid casing within depth of steel section
        B = overall width of steel flange.
Radius of gyration (r_x) about its axis parallel to plane of flange
taken as that of steel section alone.   (code 4.14.3(a))
The compression resistance P_c of the cased section is given by:
                    (code 4.14.3(b))

$$P_c = [A_g + (0.45 f_{cu}/p_y)A_c]p_c \not> P_{cs} = [A_g + (0.25 f_{cu}/p_y)A_c]p_y$$

where A_c = gross sectional area of concrete neglecting any casing in excess of
        75 mm from steel section
    A_g = gross sectional area of steel strut
    f_{cu} = characteristic 28 day cube strength of concrete encasement but
        $\leqslant 40\,\text{N/mm}^2$
    p_c = compressive strength of steel section (using r_y and r_x as defined
        above)
    p_y = design strength of steel but $\leqslant 355\,\text{N/mm}^2$
Cased members subject to axial load and moment   (code 4.14.4)

These should satisfy the following:
For capacity

$$F_c/P_{cs} + M_x/M_{cx} + M_y/M_{cy} \leqslant 1$$

For buckling resistance

$$F_c/P_c + mM_x/M_b + mM_y/M_{cy} \leqslant 1$$

where F_c = applied axial compression
    P_c = compression resistance   (code 4.14.3(b))
    P_{cs} = the short strut capacity (i.e. compression resistance
        of a cased strut of zero slenderness)
    M_x = applied moment about the major axis
    M_y = applied moment about the minor axis
    m = equivalent uniform moment factor   (code Table 18)
    M_{cx} = major axis moment capacity of steel section   (code 4.2.5)
    M_{cy} = minor axis moment capacity of steel section   (code 4.2.5)
    M_b = buckling resistance moment obtained from   (code 4.3.7.3)
                    (Section properties from code 4.14.2)

8.6.2 Column design

Column	Height m	Calculation	r_x cm	r_y cm	S_x cm³	Selected section	Code ref.
C2 Roof to 4th floor	4.0	Vertical load $125.7 + (2 \times 30.3)$ $\quad = 186.3$ kN O.W. + casing $6.7 \times 1.4 \quad = \underline{9.4}$ $\overline{195.7}$	6.51	3.68	184.3	$152 \times 152 \times 23$ UC $\quad S_y = 80.87$	
Designed as uncased to demonstrate but casing included in loading.		Eccentricity from column centreline of beam B4 $152/2 + 100 = 176$ mm $M_x = 125.7 \times 176/10^3 = 22.1$ kNm $M_y = 0$					
		Moment capacity check $M_{cx} = p_y S_x = 275 \times 184.3 \times 10^{-3}$ $\quad = 50.7$ kNm					4.2.5
		Local capacity check $F/A_g p_y + M_x/M_{cx} + M_y/M_{cy} \leqslant 1$ $195.7 \times 10^3/(29.8 \times 10^2 \times 275) + 22.1/50.7 + 0 \leqslant 1$ $0.24 + 0.44 = 0.68 \leqslant 1$ $\therefore$ local capacity adequate					4.8.3.2

Overall buckling check (simplified method)

$\lambda_y = L/r_y = 0.85 \times 4000/36.8 = 92.4$

$\lambda_x = L/r_x = 0.85 \times 4000/65.1 = 52.2$

$P_{cx} = 234\,\text{N/mm}^2$

4.8.3.2
Table 25
for structure
type. Use Table
27 (b) and (c)

$F/A_g p_c + mM_x/M_b + mM_y/p_y Z_y \leqslant 1$

$M_b = S_x p_b$. Determine p_b.

$\lambda_{LT} = 0.5L/r_y = 0.5 \times 4000/3.68 \times 10 = 54.3$

$\therefore p_b = 226\,\text{N/mm}^2$

$M_b = 184.3 \times 10^3 \times 247/10^6 = 41.65\,\text{kNm}$

$195.7 \times 10^3/(29.8 \times 10^2 \times 234) + 1 \times 22.1/41.65$

$\qquad \leqslant 1$

$0.28 + 0.53 = 0.81 \leqslant 1$

$\therefore$ Section is adequate against overall buckling

4.7.7
Table 11

239

Column design (continued)

Column	Height m	Calculation	r_x cm	r_y cm	S_x cm³	Selected section	Code ref.
C2	4.0	Loading (kN) Roof		3.87	310.1	$152 \times 152 \times 37$ UC	

3rd to 2nd floor. Designed as cased to demonstrate

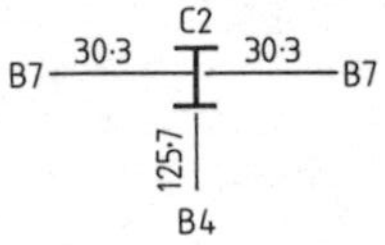

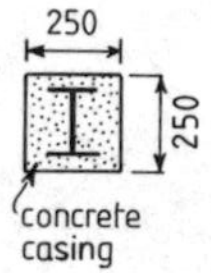

B7 ── 30·3 ── C2 ── 30·3 ── B7
125·7
B4

4th floor
(3rd floor similar)
No live load reduction as imposed floor
load $= 5\,\text{kN/m}^2$ (see note to Table 8.2)
Load on col. $224.7 + (2 \times 39.8)$
$\quad + 8.7(\text{O.W.}) = 313\,\text{kN}$
Load on column kN
From roof $= 186.3$
From 4th floor $= 313.0$
From 3rd floor $= 313.0$

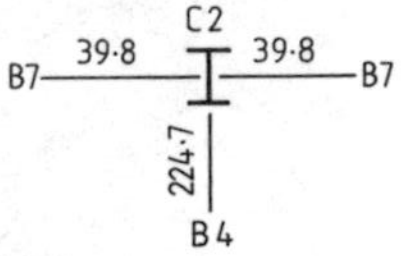

Total 812.3

concrete casing
(250 × 250)

B7 ── 39·8 ── C2 ── 39·8 ── B7
224·7
B4

To demonstrate procedure, design as a cased column
$r_y = 0.2 b_c$ where $b_c = 250\,\text{mm}$ (see sketch above)
$r_y = 0.2 \times 250 = 50\,\text{mm} \not> 0.2\,(150 + 150)$
r_x (steel section only) $= 68.4\,\text{mm}$
$\lambda = L/r_y = 0.85 \times 4000/50 = 68$
$p_c = 185\,\text{N/mm}^2$

4.14.3.

Table 27(c)

$M_x = 224.7 \times 181/10^3 = 40\,\text{kNm}$ divided equally between upper and lower columns

$M_y = 0 \quad 0.5 M_x = 20\,\text{kNm}$

Capacity check

$F_c/P_{cs} + M_x/M_{cx} + M_y/M_{cy} \leqslant 1$ 4.14.4(a)

$P_{cs} = [A_g + (0.25 f_{cu}/p_y)A_c]p_y$

$\quad = (47.4 \times 10^2 + 0.25 \times 20 \times 250 \times 250/275)275 \times 10^{-3}$

$\quad = 1616\,\text{kN}$

C2
3rd to 2nd floor

$M_{cx} = p_y S_x = 275 \times 310.1 \times 10^{-3} = 85.3\,\text{kNm}$

$812.3/1616 + 20/85.3 = 0.50 + 0.24 = 0.74 < 1$

Check buckling resistance

$F_c/P_c + mM_x/M_b + mM_y/M_{cy} \leqslant 1$ 4.14.4(b)

$P_c = (47.4 \times 10^2 + 0.45 \times 20 \times 250^2/275)185 \times 10^{-3}$

$\quad = 1256\,\text{kN} < 1616(P_{cs})$

$M_b = S_x p_b$ Determine p_b Table 19 (b)

$\lambda_{LT} = 0.5\ L/r_y = 0.5 \times 400/3.87 = 51.7$

$p_b = 235\,\text{N/mm}^2$

$M_b = 235 \times 310.1 \times 10^3/10^6 = 72.9\,\text{kNm}$

$m = 1.\ M_x = 20\,\text{kNm}$

Hence, $812.3/1256 + 20/72.9 + 0 < 1$

$\quad = 0.65 + 0.27 = 0.92 < 1$

$\therefore$ section is adequate.

Column design (continued)

Column	Height m	Calculation	r_x cm	r_y cm	S_x cm³	Selected section	Code ref.
C2 grd to 1st floor	4.0	Loading from roof = 186.3 kN 1st–4th flrs: Typical floor loading					
		B4 = 224.7					
		B7 39.8 × 2 = 79.6					
		304.3					
		Total load on column					
		Roof = 186.3					
		1st–4th. 4 × 304.3 = 1217.2					
		O.W. and casing					
		5 × 8.7 = 43.5	8.9		568.1	203 × 203 × 52 UC	
		1447.0					
		As cased column $L = 4\,\text{m}$					
		$r_y = 0.2b_c = 0.2 \times 300 = 60\,\text{mm}$					
		$L/r_y = 4000 \times 0.85/60 = 56.7$					
		$p_c = 207\,\text{N/mm}^2$					
		$M_y = 0$					
		$M_x = 224.7 \times 201/10^3 = 45.2\,\text{kNm}$					
		Divide between upper and lower					
		columns $45.2/2 = 22.6\,\text{kNm}$					
		Capacity check					
		$F_c/P_{cs} + M_x/M_{cx} + M_y/M_{cy} \leqslant 1$					
		$P_{cs} = [A_g + (0.25 \times f_{cu}/p_y)A_c]p_y$					Table 27(c)
		$= [66.4 \times 10^2 + (0.25 \times 20/275)300 \times 300]275/10^3$					
		$= 1916\,\text{kN}$					
		$M_{cx} = p_y S_x = 275 \times 568.1/10^3 = 156.3\,\text{kNm}$					
		$1447/1916 + 22.6/156.3 + 0 \leqslant 1$					
		$0.76 + 0.15 = 0.91 < 1$					

242

Check buckling resistance

$F_c/P_c + mM_x/M_b + mM_y/M_{cy} \leq 1$

$P_c = [A_g + (0.45 f_{cu}/p_y)A_c]p_c$

$= [66.4 + (0.45 \times 20/275)300 \times 300]207/10^3$

$= 1984\,kN.$

C2
grd 1st Floor

$M_b = S_x p_b$

$\lambda_{LT} = 0.5 L/r_y$

$= 0.5 \times 4000/60 = 33.3$

$x = D/T = 16.5$

$p_b = 221.6\,N/mm^2$

$M_b = 221.6 \times 568.1 \times 10^{-3} = 128\,kNm$

$1447/1984 + 22.6/128 \leq 1$

$0.73 + 0.18 = 0.91 < 1$

Section adequate

For cased section = 6.0
For steel alone = 5.16

Table 11

243

Column design (continued)

Column	Height m	Calculation	r_x cm	r_y cm	S_x cm³	Selected section	Code ref.
C9 3rd to 2nd floor	4.0	This central column has been selected to illustrate the loading requirements on columns between 3rd and 2nd floors.				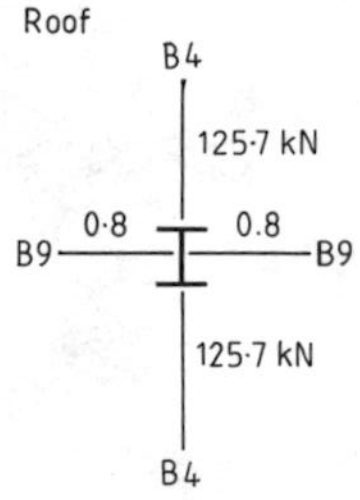	

Roof load   kN

B4 2×125.7	=	251.4
B9 2×0.8	=	1.6
O.W. + casing	=	5.8
		258.8

4th floor (3rd floor similar)

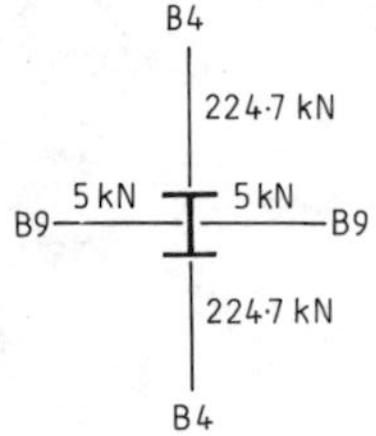

B4 2×224.7	=	449.4
B9 2×5	=	10.0
O.W. + casing	=	8.5
		467.9

Total load		
Roof	=	258.8
4th	=	467.9
3rd	=	467.9 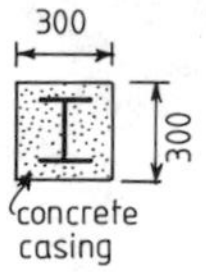
		1194.6
	say	1195

Roof
B4

125·7 kN

B9   0·8   B9

0·8

125·7 kN

B4

B4

224·7 kN

B9 — 5 kN   5 kN — B9

224·7 kN

B4

concrete casing

300

300

$203 \times 203 \times 46$ UC   Table 27 (c)

$D/T = 18.5$
$A = 58.8\,\text{cm}^2$
$S_x = 497.4\,\text{cm}^3$

Design as cased column, length 4 m
$\lambda = L/r_y = 4000 \times 0.85/(0.2 \times 300) = 56.7$
$p_c = 207.6\,\text{N/mm}^2$

r_x (steel section only) = 88.1 mm

C9
3rd to 2nd
floor

4.0

$M_x = (224.7 - 99.0)201.5/10^3 = 25.4\,\text{kNm}$
(assuming imposed load absent from one beam **B4**)
Divide equally between upper and
lower columns $25.4/2 = 12.7\,\text{kNm}$
$M_y = 0$
Capacity check
$F_c/P_{cs} + M_x/M_{cx} + M_y/M_{cy} \leqslant 1$
where $P_{cs} = [A_g + (0.25 \times f_{cu}/p_y)A_c]p_y$
$\quad = [58.8 \times 10^2 + (0.25 \times 20/275)300^2]275/10^3$
$\quad = 2067\,\text{kN}$

$M_{cx} = p_y S_x = 275 \times 497.4/10^3 = 136.8\,\text{kNm}$
$1195/2067 + 12.7/136.8 = 0.58 + 0.09 = 0.67 < 1$

Check buckling resistance
$F_c/P_c + mM/M_b + mM_y/M_{cy} \leqslant 1$
$P_c = [A_g + (0.45 \times f_{cu}/p_y)A_c]p_c$
$\quad = [58.8 \times 10^2 + (0.45 \times 20/275)300 \times 300]207.6/10^3$
$\quad = 1832\,\text{kN}$

$M_b = S_x p_b$
$\lambda_{LT} = 0.5L/r_y = 33.3$
$p_b = 273\,\text{N/mm}^2$
$M_b = 273 \times 497.4 \times 10^{-3} = 135.8\,\text{kNm}$
$1195/1832 + 1 \times 12.7/135.8 = 0.65 + 0.1$
$\quad\quad\quad\quad\quad\quad\quad\quad\quad = 0.75 < 1$

Section adequate.

Table 11

245

REFERENCES

1. Bates, W. (1975) *Medium Rise Building*. Constrado, Croydon.
2. BS648 (1964) *Schedule of Weights of Building Materials*. British Standards Institution.
3. BS6399: Part 1 (1984) *Code of Practice for Dead and Imposed Loads*. British Standards Institution.

Answers to problems

Chapter 2

2.1 3553 cm⁴.

2.2 938 cm⁴.

2.3 Angle and bolts adequate for applied load, bolt shear governs in this case with capacity of 2 No. M20 bolts grade 4.6 = 73.6 kN.

2.4 Beam satisfactory, i.e. $\bar{M}_x/M_{ax} + \bar{M}_y/kM_{ay} = 0.3 < 1.0$.

2.5 244.5 CHS × 12.5 wall thickness grade 50. Column capacity 2387 kN > 1880 kN.

2.6 (a) Average shear stress = 23 N/mm² < 213 N/mm² allowable (code 4.2.3).
(b) Elastic shear stress = 84 N/mm² < 248 N/mm² allowable (code 4.2.4).

2.7 254 × 146 × 31 UB.

Chapter 3

3.1 Base plates 220 mm square × 10 mm thick, placed symmetrically about column centreline. 6 mm fillet weld, 2 No. M20 HD bolts, grade 4.6.

3.2 550 mm sq. plt. × 30 mm thick, 4 No. M30 grade 4.6 HD bolts, 6 mm fillet weld.

3.3 Flange splice plates 12 mm thick, with 4 No. M30 grade 8.8 bolts each side of splice; end plates 12 mm thick welded with 5 mm fillet weld to column with 4 No. bolts, i.e. 2 each side of column web, M16 grade 8.8.

3.4 Splice at 3 m from one end. Flange plates 220 mm long × 150 mm × 15 mm thick + 4 No. M20 grade 8.8 bolts each side.
Web plate 120 mm long × 350 mm × 10 mm thick + 2 No. M16 grade 8.8 bolts each side.

Chapter 4

4.1 Framing angles – 2/75 × 75 × 6 angles, 4 No. M16 grade 8.8 bolts – angles to beam web, and 4 No. M16 grade 8.8 bolts angles to col. flange (2 each side).

4.2 Plate size 180 mm deep × 120 mm wide × 10 mm thick. Weld plate to top flange and to beam web with 6 mm fillet weld. Bolt plate to column flange with 3 No. M20 grade 4.6 bolts each side.

4.3 Seated connection; 200 mm × 200 mm × 16 mm angle × 250 mm long with 6 mm weld all round, and nominal 2 No. M16 grade 8.8 bolts at seat. Top angles, use 60 × 60 × 8 angle + 4 No. M16 grade 8.8 bolts.

4.4 Stiffened seat bracket; use stub of 203 × 203UC71 × 205 mm long, cut square, welded to flange of column with 6 mm fillet weld. Bolt beam to stub with nominal 4 No. M16 grade 8.8 bolts.

4.5 and 4.6 Worked solutions given with problems.

4.7 Cut haunch plate from beam section. Use 2 × 4 No. M27 HSFG bolts on 140 mm bolt gauge. End plate 25 mm thick × 250 mm wide. Add column stiffeners as follows: Use 8 No. 10 mm triangular stiffeners. (Refer to Example 4.30 as a guide to stiffener location.) Use 10 mm horizontal stiffeners lining up with top and bottom of haunch. Full strength butt welds to ends of haunch and junction of beam and end plate. 5 mm fillet welds elsewhere.

Chapter 5

5.1 24 kNm neglecting beam self weight. Recalculate using a suitable beam size.

5.2 29.5 kNm.

5.3 43.75 kNm. Hinges at A, B and midspan AB. Beam selected – 203 × 133UB25 (grade 43). Restraint provided at hinge locations (code. 5.3.5) not greater than 1840 mm from hinges; elastic restraint provided at intervals not greater than 2170 mm in 'elastic' spans.

5.4 20 kNm.

5.5 63.2 kNm.

Appendix

MISCELLANEOUS DATA

A.1 Angles of repose for contained materials. Materials should be tested for final design value.

Material	Weight in kN/m³	Angle of internal friction θ	K_1 from Rankine formula	Friction angle on steel ϕ_s	Friction angle on conc ϕ_c
Anthracite: dry & broken	8.8	27	0.376	22	27
Anthracite: pulverized aerated	6.4	20	0.490	15	20
Anthracite: pulverized compact	9.6	25	0.406	20	25
Ash: dry, compact	7.2	40	0.217	25	35
Ash: loose	6.4	30	0.333	20	30
Ash: wet saturated	11.2	20	0.490	15	20
Cement	14.4	20	0.490	—	—
Cement clinker	14.4	30–33	0.333–0.295	—	—
Cement slurry	16.5	—	—	—	—
Chalk in lumps	11.2	30	0.333	—	—
Cinders: cinder tap	20.8	—	—	—	—
Cinders: coal raw 10 mm	10.4	40	0.217	—	—
Cinders: coal gas	8.96	45–50	0.172–0.132	—	—
Coal: bituminous, dry, & broken	8.0	35	0.271	25	35
Coal: pulverized aerated	5.6	20	0.490	15	20
Coal: pulverized compact	8.8	25	0.406	20	25
Coal: Welsh	8.8	40	0.217	—	—
Coke: dry loose	4.8	30	0.333	20	20
Crushed brick	16.0	35–40	0.271–0.217	—	—
Crushed stone	20.8	35–40	0.271–0.217	—	—
Earth	19.2	26–37	0.390–0.249	—	—
Grain: barley	6.4	27	0.376	—	—
Grain: wheat	8.5	25	0.406	—	—
Gravel & sand	17.6	25–30	0.406–0.333	15	20
Limestone	17.9	35	0.271	—	—
Mud	19.2	0	1.00	—	—

Ores.					
Copper ore	28.8	35	0.271	—	—
Haematite iron ore	36.8	35	0.271	—	—
Lead ore	51.2	35	0.271	—	—
Lincolnshire iron ore	17.6	35	0.271	—	—
Magnetite iron ore	40.0	35	0.271	—	—
Manganese ore	28.8	35	0.271	—	—
Zinc ore	28.8	35	0.271	—	—
Salt	9.6	30	0.333	—	—
Sand: dry	16.0	30–35	0.333–0.271	20	30
Sand: wet	19.2	25–30	0.406–0.333	15	20
Shale	19.2	30–35	0.333–0.271	—	—
Shingle	17.6	30–40	0.333–0.217	—	—
Slag	14.4	35	0.271	—	—
Rubblestone	19.2	45	0.172	—	—

Values not shown yet to be ascertained

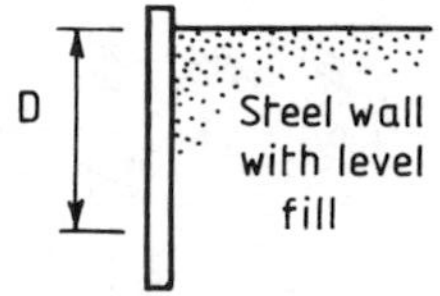

Rankine formula

$$K_1 = \frac{1 - \sin\theta}{1 + \sin\theta}$$

Pressure at depth $D = K_1\gamma D$, where γ is density.

Steel wall with level fill

D

A.2 Weights of materials

Item	Weight in kN/m³	Item	Weight in kN/m³
Acid: acetic	10.4	Douglas fir	5.1
Acid: nitric	15.0	Earth: dry, loose	11.5
Acid: sulphuric	18.0	Earth: moist, packed	15.4
Air: gas	0.013	Earth: dry, rammed	17.9
Alcohol	7.9	Eggs	10.9
Aluminium: cast	27.5	Elm	7.2
Ammonia, liquid	5.9	Fats	9.3
Asbestos	28.0	Feldspar: orthoclase, mineral	26.0
Ash: timber	6.5	Fir	5.1
Ashes: coal	7.2	Flour: loose	5.0
Asphalt for paving	23.0	Fruit: dried in cases	9.6
Ballast: brick	17.9	Gas: natural	0.006
Barley: cereal, in bulk	6.4	Glass: common	26.0
Basalt: mineral	32.0	Glass: plate	27.8
Bauxite	25.5	Glycerine in cases	8.3
Bleach in barrels	5.1	Gold: cast	193.0
Borax: mineral	18.0	Gneiss: mineral	27.0
Brass: cast	87.0	Granite: mineral	31.0
Brick	20.0	Graphite	23.0
Bronze	83.8	Gravel	19.2
Butter	8.6	Gypsum	28.0
Cedar: timber	38.0	Hay in bales, compressed	3.8
Cement: Portland powder	14.4	Hemlock: timber	5.2
Cement: Portland set	16.8	Hemp, baled	3.2
Cement: Portland in bags	13.4	Hickory: timber	8.4
Chalk	26.0	Hornblende: mineral	30.0
Cheese	4.8	Iron: cast	72.0
Chestnut: timber	6.6	Iron: wrought	76.8
Clay: kaolin	22.1	Jute	4.8
Clay: dry potters	19.1	Larch: timber	5.6
Clay: saturated	22.8	Lead: cast	113.3
Coal	17.0	Leather hides, compressed	3.7
Coffee	6.4	Limestone	28.0
Coke	4.8	Lime in barrels	8.0
Concrete	24.0	Lime mortar	16.5
Copper: cast	90.0	Linen: baled	8.0
Cotton	15.0	Linseed oil	8.8
Crockery in crates	6.4	Magnesium alloys	18.3
Cypress: timber	4.8	Magnesite: mineral	30.0
Dolomite: mineral	29.0	Masonry: granite dressed	26.4

A.2 Weights of materials (continued)

Item	Weight in kN/m^3	Item	Weight in kN/m^3
Masonry: freestone dressed	24.0	Sand: dry	16.0
Manganese	80.0	Sand: wet	20.0
Manilla: baled	4.2	Sandstone	25.0
Maple: hard	6.8	Screw nails, in packages	16.0
Marble	28.0	Sewage	11.8
Mercury	136.0	Silver: cast	106.0
Mortar	16.5	Slate	29.0
Nickel	89.0	Snow: newly fallen	1.0
Nitrogen gas	0.013	Snow: wet compact	3.2
Oak	9.6	Soaps in cases	9.0
Oats in bulk	5.1	Soda ash: in barrels	9.9
Oils in barrels	5.8	Soda: caustic in drums	14.1
Oxygen gas	0.014	Soil: see earth	—
Paper: printing	6.4	Spruce	4.6
Paper: writing	9.6	Starch	15.3
Petrol	8.7	Steel: rolled	78.4
Pine: pitch	7.2	Straw in bales, compressed	3.0
Pine: red yellow	6.4	Sugar in bags	7.2
Pine: white	4.8	Sulphur	20.7
Pitch	11.5	Sulphuric acid see acids	—
Phosphate: rock	32.0	Tar	12.0
Plaster	15.4	Tea in chests	4.0
Plaster of Paris: Loose	9.6	Terra cotta	17.9
Platinum: cast	215.0	Tinned goods in cases	9.6
Polar: timber	4.8	Tin sheet in boxes	44.5
Porphyry	29.0	Tin: cast	72.8
Potash	22.6	Tin: ore	70.0
Potatoes	7.2	Walnut	6.1
Pumice	9.0	Water: fresh	10.0
Quartz	28.0	Water: sea	10.2
Red lead: dry	21.1	Wheat: in bags	6.2
Rice in bags	9.3	Wheat: bulk	7.2
Rope in coils	5.1	Wire in coils	11.8
Rosin in barrels	7.7	Wool: compressed	7.7
Rubber	9.6	Wool: piece goods cased	4.3
Salt	7.7	Zinc: cast	72.0
Saltpetre	10.7	Zinc: ore	42.0

A.3 Cantilevers

LOADING (W = total load)	Max Shear	Max Moment Coefficient	Max Deflection Coefficient	Slope at free end of beam.
	W	1	$\frac{1}{3}$	$\emptyset = \dfrac{WL^2}{2EI}$
	W	$\frac{2}{3}$	$\frac{11}{60}$	$\emptyset = \dfrac{WL^2}{4EI}$
	W	$\frac{1}{2}$	$\frac{1}{8}$	$\emptyset = \dfrac{WL^2}{6EI}$
	W	$\frac{1}{3}$	$\frac{1}{15}$	$\emptyset = \dfrac{WL^2}{12EI}$
	W	Max moment $= W(L-b/2)$	Max defln $= \dfrac{W(8c^3-24c^2L-b^3)}{48EI}$	$\dfrac{W(b^2+12c^2)}{24EI}$
	0	Max mom= M (on whole span)	Max defln $= \dfrac{ML^2}{2EI}$	$\emptyset = \dfrac{ML}{EI}$

Max moment $=$ W×L× coefficient.

Max deflection $= \dfrac{WL^3}{EI}$ × coefficient.

$\emptyset = \dfrac{\sum a}{E.I}, \quad \delta = \dfrac{\sum ax}{E.I}$

where x=dist from free end to c.g. B.M.D

a=area of B.M.Diagram.

A.4 Simply supported beams

LOADING (W= total load)	Max shear	Max moment coefficient.	Max defln. coefficient.
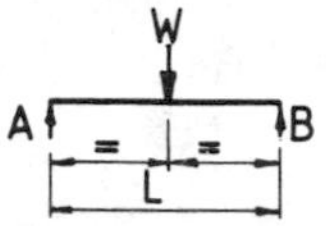	$\dfrac{W}{2}$	$\dfrac{1}{4}$	$\dfrac{1}{48}$
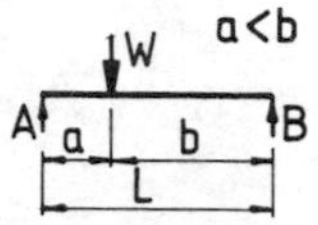 $a<b$	$R_A= \dfrac{Wb}{L}$ $R_B= \dfrac{Wa}{L}$	Max moment = $\dfrac{Wab}{L}$	δ max = $\dfrac{Wab}{27EIL}(2a+b)\sqrt{3b(2a+b)}$
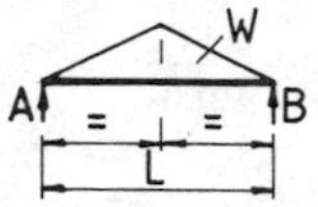	$\dfrac{W}{2}$	$\dfrac{1}{6}$	$\dfrac{1}{60}$
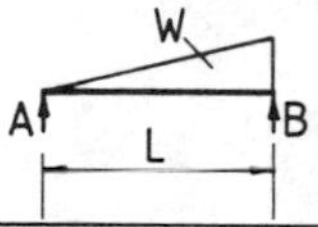	$R_B=\dfrac{2W}{3}$ $R_A = \dfrac{W}{3}$	$\dfrac{1}{7.8125}$	$\dfrac{1}{76.687}$
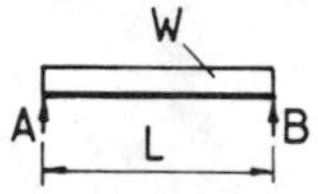	$\dfrac{W}{2}$	$\dfrac{1}{8}$	$\dfrac{5}{384}$ $(= 1/76.8)$
$a<b$	$R_A =W\left[1-\dfrac{a}{2L}\right]$ $R_B = \dfrac{Wa}{2L}$	M max = $\dfrac{Wa}{2}\left[1-\dfrac{a}{2L}\right]^2$	at centre, $\delta=$ $\dfrac{W}{384EI}(8L^3-4a^2L+a^3)$
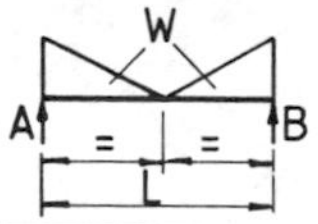	$\dfrac{W}{2}$	$\dfrac{1}{12}$	$\dfrac{3}{320}$ $(= 1/106.67)$
$a>b$ 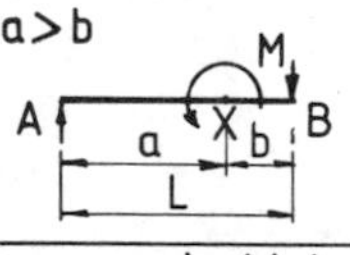	$R_A=R_B= \dfrac{M}{L}$	$M_{xA} = \dfrac{M.a}{L}$ $M_{xB} = \dfrac{M.b}{L}$	when a=L, δ centre = $\dfrac{ML^2}{16EI}$

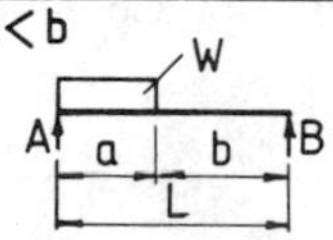

Max moment = W×L× coefficient.   Max defln = (WL^3/EI) × coefficient.

Max deflection for a simply supported beam occurs within 0.0774L of beam centre, the exception to this rule is where moment is applied, as shown above.

A.5(a) Propped cantilevers

LOADING (W= total load)	Max shear	Max moment coefficient.	Max defln. coefficient.
A ⊥ W↓ ×⎯ B, L	$R_A = \dfrac{11W}{16}$	$M_A = \dfrac{3}{16}$ $\left[M_x = \dfrac{5}{32}\right]$	$\dfrac{1}{107.296}$
A ⊥ W B, L	$R_A = \dfrac{21W}{32}$	$\dfrac{5}{32}$	$\dfrac{1}{137.552}$
A ⊥ W B, L	$R_A = \dfrac{5W}{8}$	$M = \dfrac{1}{8}$ M span $= \dfrac{9}{128}$	$\dfrac{1}{185}$

Max moment = W×L×coefficient.   Max deflection = (WL^3/EI) × coefficient

A.5(b) Properties of a parabola (useful for calculating moments and deflections using the Area–Moment method)

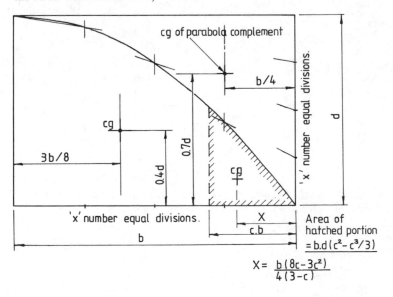

Area of hatched portion $= b.d\,(c^2 - c^3/3)$

$$X = \frac{b\,(8c - 3c^2)}{4\,(3 - c)}$$

A.6 Fixed ended beams

LOADING (W = total load)	Max shear	Max moment coefficient	Max defln coefficient
	$\dfrac{W}{2}$	$M = \dfrac{1}{8}$ $\left[M \text{ mid span} = \dfrac{1}{8}\right]$	$\dfrac{1}{192}$
	$R_A =$ $\dfrac{W b^2 (L+2a)}{L^3}$	$M_A = \dfrac{Wab^2}{L^2}$ $\left[M_x = \dfrac{2Wa^2b^2}{L^3}\right]$	b $\;\;\delta$ max at dist x 0.5 | 1/192 | 0.5 0.4 | 1/233.6 | 0.483 0.3 | 1/349 | 0.478 0.2 | 1/694.8 | 0.481 0.1 | 1/2549.4 | 0.489
	$\dfrac{W}{2}$	$M_A = \dfrac{5}{48}$ $\left[M \text{ mid span} = \dfrac{1}{16}\right]$	$\dfrac{1}{274.286}$
	$\dfrac{W}{2}$	$M_A = \dfrac{1}{12}$ $\left[M \text{ mid span} = \dfrac{1}{24}\right]$	$\dfrac{1}{384}$
	$R_A =$ $W\left[1 - \dfrac{a}{2L}\right]$ $+ \dfrac{M_A - M_B}{L}$	with b = 0.5L, $M\max = M_B = \dfrac{1}{8.72}$	b $\;\;\delta$ max at dist x 0.5 | $\dfrac{1}{782.2}$ | 0.444
	$R_A = R_B =$ $\dfrac{M_{Rx} + M_{xA}}{a}$	$M_A = \dfrac{M.b}{L^2}(3a-L)$	b $\;\;\delta$ max at dist x. 0.5 | $\dfrac{ML^2}{288EI}$ | 0.333
Max moment = W × L × coefficient. Max deflection = (WL³/EI) × coefficient.			

A.7 Pitched roof portal frames – Definitions and frame constants

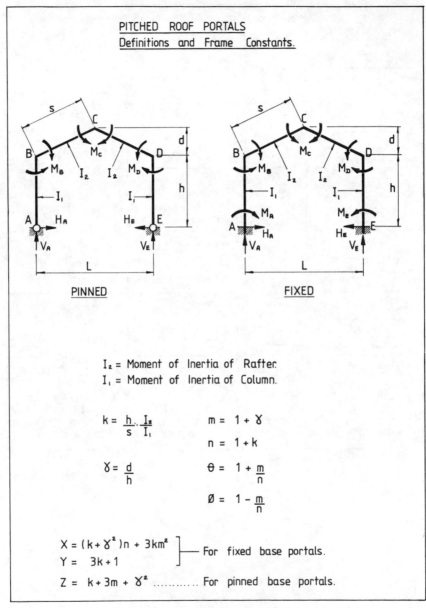

PITCHED ROOF PORTALS
Definitions and Frame Constants.

PINNED

FIXED

I_2 = Moment of Inertia of Rafter.
I_1 = Moment of Inertia of Column.

$$k = \frac{h}{s} \cdot \frac{I_2}{I_1}$$

$$m = 1 + \gamma$$

$$n = 1 + k$$

$$\gamma = \frac{d}{h}$$

$$\theta = 1 + \frac{m}{n}$$

$$\emptyset = 1 - \frac{m}{n}$$

$$X = (k + \gamma^2)n + 3km^2$$
$$Y = 3k + 1$$

⎱ For fixed base portals.

$$Z = k + 3m + \gamma^2 \quad \dots\dots\dots \text{ For pinned base portals.}$$

(A frame formulae. Tables reproduced with the permission of Quasi Arc Co. Ltd).

A.8 Pitched roof, pinned base portal frames

LOADING	F	Q	M_B	M_C	M_D
	$\dfrac{1}{8}$	$\dfrac{5m+3}{32Z}$	$WL(-Q)$	$WL(F-mQ)$	$WL(-Q)$
	$\dfrac{1}{8}$	$\dfrac{5m+3}{32Z}$	$WL(-Q)$	$WL(F-mQ)$	$WL(-Q)$
	$\dfrac{1}{2}$	$\dfrac{\gamma(3m+1)}{16Z}$	$Wh(F+Q)$	$Wh\left[-\dfrac{\gamma F}{2}+mQ\right]$	$Wh(-F+Q)$
	$\dfrac{1}{2}$	$\dfrac{6m+5n+1}{16Z}$	$Wh(F-Q)$	$Wh\left[\dfrac{F}{2}-mQ\right]$	$Wh(-Q)$
$\propto=\dfrac{a}{L}$	$\dfrac{1}{2}$	$\dfrac{\gamma(3-4\alpha^2)+6(1-\alpha)}{4Z}$	$Wa(-Q)$	$Wa(F-mQ)$	$Wa(-Q)$
$\beta=\dfrac{b}{d}$	$\dfrac{1}{2}$	$\dfrac{\beta^2\gamma(3m-\beta\gamma)}{4Z}$	$Wh(F+Q)$	$Wh(-\gamma\beta F+mQ)$	$Wh(-F+Q)$
$\propto=\dfrac{a}{h}$	1	$\dfrac{3m+2n+1+k(1-\alpha^2)}{4Z}$	$Wa(F-Q)$	$Wa\left[\dfrac{F}{2}-mQ\right]$	$Wa(-Q)$
$\propto=\dfrac{a}{h}$	1	$\dfrac{3(m+n-k\alpha^2)}{4Z}$	$Wc(F-Q)$	$Wc\left[\dfrac{F}{2}-mQ\right]$	$Wc(-Q)$

A.9 Pitched roof, fixed base portal frames

LOADING	F	G	Q	R	M_A
	$\dfrac{1}{8}$	$\dfrac{1}{12n}$	$\dfrac{4km+\delta m}{16X}$	—	$WL(-G+Q\theta)$
	$\dfrac{1}{4}$	$\dfrac{3k+1}{24n}$	$\dfrac{4km+\delta m}{16X}$	$\dfrac{3(4k+1)}{32Y}$	$WL(-F+G+Q\theta+R)$
	$\dfrac{\delta}{2}$	$\dfrac{3km+\delta}{12n}$	$\big[2km(3m-\delta)+n(2k+\delta^2)\big]\div 8X$	$\dfrac{3(4km+\delta)}{16Y}$	$Wh(-1-F+G+Q\theta+R)$
	$\dfrac{1}{2}$	$\dfrac{k}{12n}$	$\dfrac{k(2m+n)}{8X}$	$\dfrac{k}{4Y}$	$Wh(-F+G+Q\theta+R)$
$\alpha=\dfrac{a}{L}$	1	$\dfrac{k+\alpha}{2n}$	$\big[3m(k+\alpha)-(3\alpha+4\delta\alpha^2)n\big]\div 2X$	$\dfrac{3(k+\alpha)-2\alpha^2}{2Y}$	$Wa(-F+G+Q\theta+R)$
$\alpha=\dfrac{a}{d}$	$\alpha\delta$	$\dfrac{\alpha^2\delta+k(1+2\alpha\delta)}{4n}$	$\big[3mk(2\alpha\delta+1)+nk-\alpha^2\delta(3n-3m+2\delta n)\big]\div X$	$\big[3k(1+2\alpha\delta)+\alpha^2\delta(3-\alpha)\big]\div 4Y$	$Wh(-1-F+G+Q\theta+R)$
$\alpha=\dfrac{a}{h}$	1	$\dfrac{\alpha k}{4n}$	$\big[\alpha k[3(m+n)-2\alpha n]\big]\div 4X$	$\dfrac{3\alpha k}{4Y}$	$Wa(-F+G+Q\theta+R)$
$\alpha=a/h$ $\beta=b/h$	1	$\dfrac{\alpha k}{2n}$	$\dfrac{3\alpha k(\beta n+m)}{2X}$	$\dfrac{3\alpha k}{2Y}$	$Wc(-F+G+Q\theta+R)$

A.10 Pitched roof, fixed base portal frames

M_B	M_C	M_D	M_E	Sketch of B.M. Diagram.
WL(-G-Qφ)	WL[F-G-Q(φ+2δ)]	WL(-G-Qφ)	WL(-G+Qθ)	
WL(-F+G-Qφ+R)	WL(G-Q(φ+2δ))	WL(G-Qφ-R)	WL(G+Qφ-R)	
Wh(-F+G-Qθ+R)	Wh(G-Q[φ+2δ])	Wh(G-Qφ-R)	Wh(G+Qθ-R)	
Wh(G-Qφ+R)	Wh(G-Q[φ+2δ])	Wh[G-Qφ-R]	Wh(G+Qθ-R)	
Wa(-F+G-Qφ+R)	Wa(G-Q(φ+2δ))	Wa(G-Qφ-R)	Wa(G+Qθ-R)	
Wh(-F+G-Qφ+R)	Wh(G-Q(φ+2δ))	Wh(G-Qφ-R)	Wh(G+Qθ-R)	
Wa(G-Qφ+R)	Wa(G-Q(φ+2δ))	Wa(G-Qφ-R)	Wa(G+Qθ-R)	
Wc(G-Qφ+R)	Wc(G-Q(φ+2δ))	Wc(G-Qφ-R)	Wc(G+Qθ-R)	

A.11 Rectangular portal frames – Definitions and frame constants

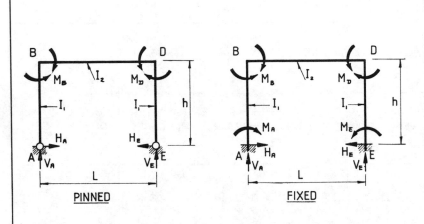

PINNED   FIXED

The rectangular portal is a special case of the pitched portal in which the dimension $d = 0$. The formulae for pitched roof portals can therefore be used by putting $d = 0$ and $s = 0.5L$.

The value of the frame constants is then as follows :–

I_1 = Moment of Inertia of Column.

I_2 = Moment of Inertia of Rafter.

$$k = \frac{2h\,I_2}{L\,I_1} \qquad\qquad m = 1$$
$$n = 1 + k$$
$$\gamma = 0 \qquad\qquad \theta = 1 + \frac{m}{n}$$
$$\emptyset = 1 - \frac{m}{n}$$

$$\left.\begin{array}{l} X = k(n+3) \\ Y = 3k+1 \end{array}\right\} \text{— For fixed base portals.}$$

$$Z = k+3 \quad\ldots\ldots\ldots\ldots \text{For pinned base portals.}$$

A.12 Portal frames – Calculation of horizontal and vertical reactions

NOTES

1   Moments causing tension on the inside faces of the portal frame are denoted positive.

2   Moments causing tension on the outside faces of the frame are denoted negative.

Horizontal Reactions

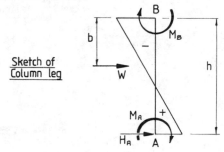

Sketch of Column leg

taking moments about B,

$$H_A = \frac{1}{h}\,(M_A - M_B) - \sum \frac{W.b}{h} \qquad \ldots\ldots\text{take note of moment signs.}$$

Vertical Reactions

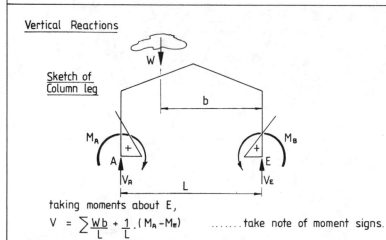

Sketch of Column leg

taking moments about E,

$$V = \sum \frac{Wb}{L} + \frac{1}{L}\,(M_A - M_E) \qquad \ldots\ldots\text{take note of moment signs.}$$

A.13 Bolt erection clearances

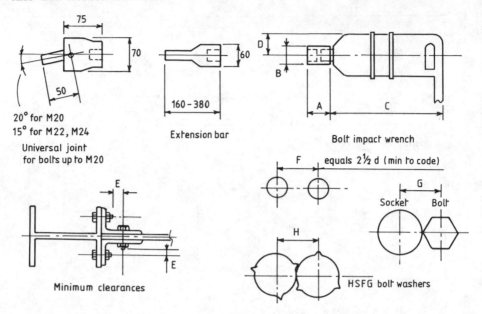

20° for M20
15° for M22, M24

Universal joint
for bolts up to M20

Extension bar

Bolt impact wrench

F   equals 2½ d (min to code)

Socket   Bolt

Minimum clearances

HSFG bolt washers

Bolt size	Sockets		Wrench			Min clearance	Bolt to bolt dimensions			Bolt size
	A	B	Type	C	D	E	F	G	H	
M16	67	45				32	40	37	38	M16
M20	76	57	Light	340–360	54	32	50	44	47	M20
M22	83	64				35	55	50	53	M22
M24	89	67				37	60	56	57	M24
M27	95	73				40	68	60	63	M27
M30	102	79	Heavy	375–440	64	43	75	66	69	M30
M36	111	108				57	90	85	89	M36

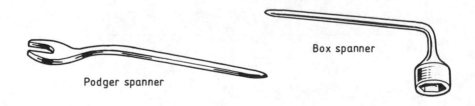

Podger spanner

Box spanner

A.14 Holding down bolts

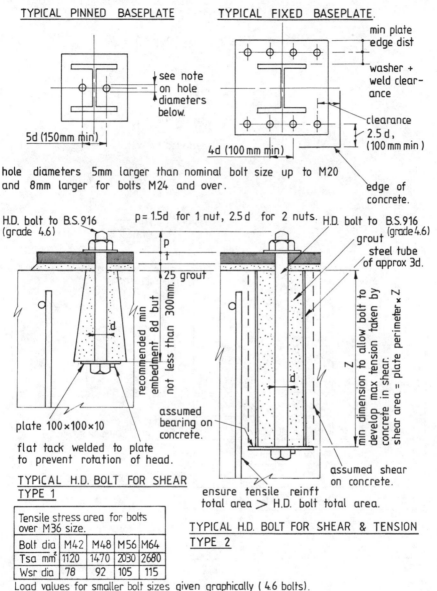

TYPICAL PINNED BASEPLATE

5d (150mm min)

see note
on hole
diameters
below.

TYPICAL FIXED BASEPLATE.

min plate
edge dist

washer +
weld clear-
ance

clearance
2.5 d,
(100 mm min)

4d (100 mm min)

hole diameters 5mm larger than nominal bolt size up to M20
and 8mm larger for bolts M24 and over.

edge of
concrete.

H.D. bolt to B.S.916
(grade 4.6)

p= 1.5d for 1 nut, 2.5 d for 2 nuts.

H.D. bolt to B.S.916
(grade 4.6)

p

t

25 grout

recommended min
embedment 8d but
not less than 300mm.

d

plate 100×100×10

flat tack welded to plate
to prevent rotation of head.

assumed
bearing on
concrete.

grout

steel tube
of approx 3d.

Z min dimension to allow bolt to
develop max tension taken by
concrete in shear.
shear area = plate perimeter × Z

d

assumed shear
on concrete.

ensure tensile reinft
total area > H.D. bolt total area.

TYPICAL H.D. BOLT FOR SHEAR
TYPE 1

TYPICAL H.D. BOLT FOR SHEAR & TENSION
TYPE 2

Tensile stress area for bolts over M36 size.				
Bolt dia	M42	M48	M56	M64
Tsa mm²	1120	1470	2030	2680
Wsr dia	78	92	105	115

Load values for smaller bolt sizes given graphically (4.6 bolts).

A.14 Holding down bolts (continued)

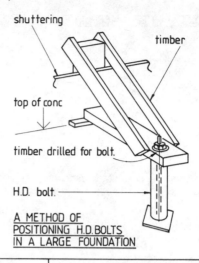

shuttering

timber

top of conc

timber drilled for bolt.

H.D. bolt.

<u>A METHOD OF</u>
<u>POSITIONING H.D. BOLTS</u>
<u>IN A LARGE FOUNDATION</u>

H.D BOLTS. TYPE 2.					area considered for bearing stress at 5.3 N/mm²
BOLT DIA	Max bolt tension kN	Tube dia × t mm×mm	Min end plt size mm	Dimn Z mm	
M16	28.	48.3×3.2	90sq×15	360	
M20	44.	60.3×3.2	110sq×15	450	
M24	63.5	76.1×3.2	130sq×20	540	
M30	101.	88.9×3.2	160sq×25	690	product of plate perimeter & dimn Z considered for shear stress at 0.23 N/mm².
M36	147.	114.3×3.6	200sq×30	810	
M42	201.	139.7×5.0	250sq×35	870	PLAN ON END PLATE
M48	264.	139.7×5.0	260sq×40	1110	
M56	365.	168.3×5.0	310sq×45	1290	
M64	482.	193.7×5.4	350sq×50	1500	

A.15 Electric overhead travelling cranes: single girder type. Medium duty

Flush construction

Cut-down construction

Normal construction

Span

Carriage length

Wheel c/c

A.15 Electric overhead travelling cranes: single type. Medium duty (continued)

Load capacity t	Span m	Wt of crane + hoist kN	Wt of hoist kN	Simultaneous Max individual wheel load kN	Min individual wheel load kN	End carriage Length m	Wheel c/c m	Clearances in metres A	B	Rail to ceiling C	Hook to ceiling D	Size of crane rail kg/m
1	4.57	13.60	2.10	9.20	2.60	2.34	1.98	0.79	0.79	0.48	0.88	12.9
	6.09	14.20	2.10	9.40	2.70	2.34	1.98	0.79	0.79	0.48	0.88	12.9
	10.67	19.00	2.10	10.50	4.00	2.34	1.98	0.79	0.79	0.48	1.03	12.9
	12.19	21.20	2.10	11.10	4.50	2.34	1.98	0.79	0.79	0.48	1.03	12.9
	15.24	27.00	2.10	12.90	5.60	2.94	2.59	0.86	0.79	0.48	1.10	12.9
	18.29	43.70	2.10	16.60	10.25	2.94	2.59	0.91	0.79	0.57*	1.30	12.9
2	4.57	15.30	2.90	15.20	2.45	2.34	1.98	0.79	0.79	0.48	0.95	12.9
	6.09	16.10	2.90	15.40	2.65	2.34	1.98	0.79	0.79	0.48	0.95	12.9
	10.67	21.00	2.90	16.70	3.80	2.34	1.98	0.79	0.79	0.48	1.10	12.9
	12.19	24.80	2.90	17.70	4.70	2.34	1.98	0.79	0.79	0.48	1.18	12.9
	15.24	31.90	2.90	20.40	5.55	2.94	2.59	0.86	0.79	0.48	1.26	12.9
	18.29	43.90	2.90	21.50	10.45	2.94	2.59	0.91	0.79	0.57*	1.37	12.9

* flush construction values

Note: Wheel loads do not include
allowance for impact
Dimension E varies, $\cong 200$ in general
*Estimated value

3	4.57	18.70	4.50	20.80	3.55	2.34	1.98	0.86	0.86	0.48	1.17	12.9
	6.09	19.90	4.50	21.10	3.85	2.34	1.98	0.86	0.86	0.48	1.17	12.9
	10.67	25.10	4.50	22.90	4.65	2.34	1.98	0.86	0.86	0.48	1.22	12.9
	12.19	30.10	4.50	24.20	5.85	2.34	1.98	0.86	0.86	0.48	1.30	12.9
	15.24	37.50	4.50	26.40	7.35	2.94	2.59	0.86	0.86	0.48	1.37	12.9
	18.29	53.40	4.50	29.10	12.60	2.94	2.59	0.91	0.86	0.64†	1.56	12.9
5	4.57	21.20	6.50	30.80	4.80	2.34	1.98	0.86	0.99	0.48	1.40	27.78
	6.09	22.40	6.50	31.10	5.10	2.34	1.98	0.86	0.99	0.48	1.40	27.78
	10.67	30.40	6.50	34.70	5.50	2.34	1.98	0.86	0.99	0.48	1.47	27.78
	12.19	35.50	6.50	36.50	6.25	2.34	1.98	0.86	0.99	0.48	1.55	27.78
	15.24	42.40	6.50	39.40	6.80	2.94	2.59	0.86	0.99	0.48	1.55	27.78
	18.29	65.30	6.50	42.10	15.55	2.94	2.59	0.91	0.99	0.76‡	1.74	27.78
7.5	4.57	27.40	11.20	36.50	14.70	2.94	2.59	1.09	1.30	0.76‡	1.80	27.78
	6.09	29.60	11.20	39.70	10.20	2.94	2.59	1.09	1.30	0.76‡	1.80	27.78
	10.67	40.10	11.20	45.80	11.75	2.94	2.59	1.09	1.30	0.76‡	1.80	27.78
	12.19	42.90	11.20	47.60	11.35	2.94	2.59	1.09	1.30	0.76‡	1.80	27.78
	15.24	64.40	11.20	53.80	15.90	2.94	2.59	1.14	1.04	0.76‡	1.75	27.78
	18.29	77.70	11.20	57.50	18.85	2.94	2.59	1.14	1.04	0.83‡	1.83	27.78
10	4.57	28.90	11.20	46.30	18.15	2.94	2.59	1.09	1.30	0.76‡	1.80	27.78
	6.09	31.70	11.20	50.40	15.45	2.94	2.59	1.09	1.30	0.76‡	1.80	27.78
	10.67	46.50	11.20	59.40	13.85	2.94	2.59	1.09	1.30	0.76‡	1.80	27.78
	12.19	50.20	11.20	61.00	14.10	2.94	2.59	1.09	1.30	0.76‡	1.80	27.78
	15.24	73.70	11.20	67.80	19.05	2.94	2.59	1.14	1.04	0.83‡	1.83	27.78
	16.76	79.00	11.20	69.50	20.00	2.94	2.59	1.14	1.04	0.83‡	1.83	34.72

† Estimated value, not in manufacturers range
‡ Cut-down construction values
Reproduced by kind permission of Herbert Morris Cranes Ltd. Values given are approximate; for current values see manufacturers tables.

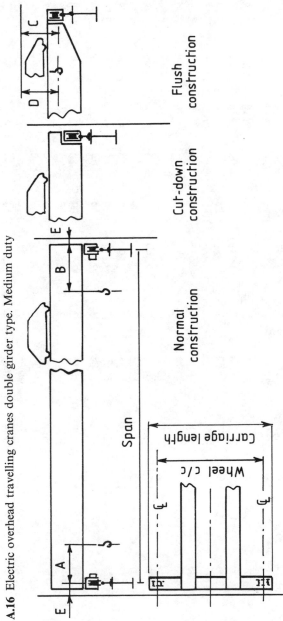

A.16 Electric overhead travelling cranes double girder type. Medium duty

A.10 (continued)

Load capacity t	Span m	Wt of crane +crab kN	Wt of crab only kN	Simultaneous Max individual wheel load kN	Min individual wheel load kN	End carriage Length m	Wheel c/c m	Clearances in metres A	B	Rail to ceiling C	Hook to ceiling D	Crane rail kg/m
1	4.57	17.50	4.0	10.0	3.75	2.34	1.98	0.76	0.76	0.90	0.65	12.9
	6.09	19.00	4.0	10.5	4.00	2.34	1.98	0.76	0.76	0.90	0.65	12.9
	10.67	31.50	4.0	13.6	7.15	2.34	1.98	0.76	0.76	0.90	0.65	12.9
	19.81	66.00	4.0	23.0	15.00	3.42	3.05	0.76	0.76	0.98	0.65	12.9
	24.38	91.00	4.0	29.0	21.50	4.02	3.65	0.76	0.76	1.03	0.65	12.9
	30.48	126.00	4.0	38.0	30.00	4.80	4.42	0.76	0.76	1.03	0.65	12.9
2	4.57	18.00	4.0	13.4	5.60	2.34	1.98	0.76	0.76	0.90	0.65	12.9
	6.09	21.00	4.0	15.5	5.00	2.34	1.98	0.76	0.76	0.90	0.65	12.9
	10.67	32.50	4.0	19.4	6.85	2.34	1.98	0.76	0.76	0.90	0.65	12.9
	19.81	66.00	4.0	28.0	15.00	3.42	3.05	0.76	0.76	0.98	0.65	12.9
	24.38	91.00	4.0	34.0	21.50	4.04	3.66	0.76	0.76	1.03	0.65	12.9
	30.48	126.00	4.0	43.0	30.00	4.78	4.42	0.76	0.76	1.03	0.65	12.9
3	4.57	19.50	6.0	18.5	6.25	2.34	1.98	0.76	0.76	0.98	0.74	12.9
	6.09	23.00	6.0	20.2	6.30	2.34	1.98	0.76	0.76	0.98	0.74	12.9
	10.67	37.50	6.0	26.5	7.25	2.34	1.98	0.76	0.76	0.98	0.74	12.9
	19.81	68.00	6.0	34.0	15.00	3.42	3.05	0.76	0.76	1.05	0.74	12.9
	24.38	93.00	6.0	40.0	21.50	4.04	3.66	0.76	0.76	1.11	0.74	12.9
	30.48	128.00	6.0	49.0	30.00	4.78	4.42	0.76	0.76	1.11	0.74	12.9
5	4.57	24.00	9.0	28.0	9.00	2.34	1.98	0.76	0.77	1.12	0.98	27.78
	6.09	29.00	9.0	31.5	8.00	2.34	1.98	0.76	0.77	1.12	0.98	27.78
	10.67	41.50	9.0	36.5	9.25	2.34	1.98	0.76	0.77	1.12	0.98	27.78
	19.81	79.00	9.0	46.0	18.50	3.42	3.05	0.76	0.77	1.19	0.98	27.78
	24.39	103.00	9.0	54.0	22.50	4.04	3.66	0.76	0.77	1.24	0.98	27.78
	30.48	143.00	9.0	62.0	34.50	4.88	4.42	0.76	0.77	1.32	0.98	27.78

FLUSH CONSTN. VALUES

Note: Wheel loads do not include allowance for impact.
Dimension E varies, $\cong 200$ in general

271

A.16 (continued)

Load capacity t	Span m	Wt of crane + crab kN	Wt of crab only kN	Simultaneous Max individual wheel load kN	Min individual wheel load kN	End carriage Length m	Wheel c/c m	Clearances in metres A	B	Rail to ceiling C	Hook to ceiling D	Crane rail kg/m
10	4.57	29.50	15.0	51.0	13.75	2.94	2.59	0.76	0.81	1.04	1.05	27.78
	6.09	34.50	15.0	55.0	12.25	2.94	2.59	0.76	0.81	1.04	1.05	27.78
	10.67	52.50	15.0	62.6	13.65	2.94	2.59	0.76	0.81	1.04	1.05	27.78
	19.81	104.00	15.0	81.0	21.00	3.82	3.35	0.76	0.81	1.16	1.05	27.78
	24.38	139.00	15.0	90.0	29.50	4.12	3.66	0.76	0.81	1.16	1.05	27.78
	30.48	190.00	15.0	103.0	42.00	4.88	4.42	0.76	0.81	1.27	1.05	34.72
15	6.09	41.00	18.0	79.9	15.60	3.24	2.74	0.88	0.88	1.28	1.80	27.78
	12.19	71.70	18.0	92.1	18.75	3.27	2.74	0.88	0.88	1.28	1.80	27.78
	16.76	114.40	18.0	103.5	28.70	3.81	3.20	0.88	0.88	1.28	1.80	27.78
	21.79	130.70	18.0	108.4	31.95	4.27	3.66	0.88	0.88	1.42	1.80	27.78
	25.60	183.50	18.0	122.8	43.95	4.57	3.96	0.88	0.88	1.44	1.80	34.72
	30.48	236.50	18.0	135.8	57.45	5.07	4.42	0.88	0.88	1.44	1.80	34.72
20	6.09	55.50	18.5	105.9	21.85	3.81	3.20	0.78	0.95	1.28	1.80	27.78
	12.19	90.70	18.5	121.0	24.35	3.81	3.20	0.78	0.95	1.28	1.80	27.78
	16.76	109.60	18.5	127.5	27.30	4.27	3.62	0.78	0.95	1.28	1.80	34.72
	21.95	171.90	18.5	143.6	42.35	4.50	3.81	0.78	0.95	1.42	1.80	34.72
	25.60	215.80	18.5	155.4	52.50	4.68	3.96	0.78	0.95	1.44	1.80	34.72
	30.48	265.50	18.5	168.3	64.45	5.14	4.42	0.78	0.95	1.44	1.80	34.72

FLUSH CONSTRUCTION VALUES

Note: Wheel loads do not include allowance for impact.
Reproduced by kind permission of Herbert Morris Cranes Ltd. Values given are approximate; for current values see manufacturers tables.

A.17 Geared travel hoists (by kind permission of Herbert Morris Limited)

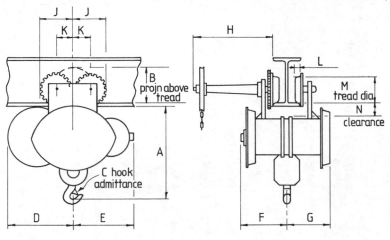

Capacity t	Standard lift m	High lift m	Weights kg							
			SL unit	HL unit	A	B	C	D	E	F*
1.0	8.23	15.24	188	222	451	121	25	330	343	514
2.0	8.53	15.24	300	340	521	175	29	406	406	397
3.0	7.62	15.24	494	538	635	178	38	470	476	460
5.0	8.53	15.24	735	815	800	197	54	610	552	508
10.0	12.19	18.29	1502	1570	1016	305	73	610	692	787

G*	H	J	K	L	M	N	Min track radius	Min track depth	Track width
518	584	165	89	35	133	70	1500	203	102–254
546	610	197	114	38	152	54	1800	203	102–254
667	762	216	117	44	165	45	1800	254	127–279
721	813	216	124	49	184	45	1800	254	127–305
1041	787	333	178	62	222	35	2700	305	152–483

Dimensions are in mm unless otherwise noted. *Largest (HL) dimension given.

A.18 Cladding data

Sheeting	Manufacturer	Sheet thickness mm	Sheet depth mm	Max centres of purlins m	Max* U.D.L. at this span kN/m²	Max centres of side rails m	Max* U.D.L. at this span kN/m²	Weight as laid horiz. kN/m²	Material finish and colour
Cladding profile 105	Brown & Clegg Ltd. Edinburgh + Adam G. Brown & Co. Ltd Glasgow	0.55	22	1.69	0.90	2.0	0.85	0.048	Profiled steel sheet available in galvanised finish or plastisol coated in a range of colours
		0.60	22	1.76		2.0	0.92	0.052	
		0.70	22	1.91		2.0	1.10	0.061	
Cladding profile 148		0.55	38	2.54		2.75	0.76	0.052	
		0.60	38	2.64		3.0	0.67	0.056	
		0.70	38	2.85		3.25	0.67	0.067	
		0.80	38	3.05		3.25	0.78	0.076	
		1.0	38	3.37		3.35	0.70	0.094	

Note: All information on this page is intended as a guide only. For more details contact manufacturer
Use movement joint in cladding when length of roofing exceeds 50 metres.
Two span condition.
* Indicates factored loading i.e. DL × 1.4 + (Wind & Imposed load) × 1.2

A.19 Floor plate simply supported on two sides

COLLAPSE LOADS
UDL in kN/m² for various spans for plate of $f_y = 240\,\text{N/mm}^2$.

Plate thickness mm	Span in metres						
	0.6	0.7	0.8	0.9	1.0	1.1	1.2
4.5	27.00	19.84	15.19	12.00	9.72	8.03	6.75
5.0	33.33	24.49	18.75	14.81	12.00	9.92	8.33
6.0	48.00	35.27	27.00	21.33	17.28	14.28	12.00
8.0	85.33	62.69	48.00	37.93	30.72	25.39	21.33
10.0	133.33	97.96	75.00	59.26	48.00	39.67	33.33
12.0	192.00	141.06	108.00	85.33	69.12	57.12	48.00
12.5	208.33	153.06	117.19	92.59	75.00	61.98	52.08

For plate fully fixed on two opposite sides, collapse loading is twice the above UDL values.

GENERAL FORMULAE FOR PLATE
where t = plate thickness, and l = span

Support type	Collapse UDL(w) in kN/m²	Elastic deflection δ
Simply supported on 2 sides.	$f_y = 240\,\text{N/mm}^2$ $0.48\,(t/l)^2$ mm m	(use consistent units) $5wl^4/32Et^3$
Fully fixed 2 sides	$0.96\,(t/l)^2$ mm m	$wl^4/32Et^3$
Simply supported 4 sides	$k(t/l)^2 \begin{pmatrix} t\,\text{in mm} \\ l\,\text{in m} \end{pmatrix}$ $k = \dfrac{1.44\,\alpha^2}{[\sqrt{1+3\alpha^2}-1]^2}$ where $\alpha \geqslant 1.0$	$\dfrac{w.k.\ 3(\alpha l)^4.(1-v^2)}{\pi^2.E.t^3\,[1+\alpha^2]^2}$ $k = 0.5634 + 0.1284\sqrt{1/\alpha}$ v = Poissons ratio (0.3 for stl) &E for steel = $206\,\text{kN/mm}^2$

Formulae by A. Haji

Plate for bunkers and hoppers can be economically designed as fixed on two opposite sides, spanning onto suitable support members. Avoid designing plate as fixed on four sides (where possible) because of its effect on welding costs.

A.20 Floor plate simply supported on four sides.
COLLAPSE LOADS

UDL in kN/m^2 for various spans for plate of $f_y = 240$ N/mm^2 based on a yield line theory calculation by A. Haji.

Plate thickness mm	Plate width m	0.6	0.7	0.8	0.9	1.0	1.1	1.2
4.5	0.6	81.0	70.04	62.61	57.27	53.28	50.18	47.72
	0.7		59.51	52.42	47.37	43.61	40.72	38.44
	0.8			45.56	40.71	37.12	34.38	32.21
	0.9				36.00	32.53	29.89	27.82
	1.0					29.16	26.60	24.60
	1.1						24.10	22.15
	1.2							20.25
5.0	0.6	100.00	86.47	77.29	70.71	65.77	61.96	58.92
	0.7		73.47	64.71	58.48	53.84	50.27	47.45
	0.8			56.25	50.26	45.83	42.44	39.77
	0.9				44.44	40.17	36.91	34.35
	1.0					36.00	32.84	30.37
	1.1						29.75	27.35
	1.2							25.00
6.0	0.6	144.00	124.52	111.30	101.82	94.71	89.22	84.84
	0.7		105.80	93.18	84.21	77.53	72.39	68.33
	0.8			81.00	72.37	65.99	61.11	57.27
	0.9				64.00	57.84	53.14	49.47
	1.0					51.84	47.29	43.73
	1.1						42.84	39.38
	1.2							36.00
8.0	0.6	256.00	221.36	197.86	181.00	168.38	158.61	150.83
	0.7		188.08	165.66	149.70	137.83	128.70	121.48
	0.8			144.00	128.66	117.32	108.64	101.82
	0.9				113.78	102.82	94.48	87.94
	1.0					92.16	84.07	77.75
	1.1						76.17	70.02
	1.2							64.00
10.00	0.6	400.00	345.88	309.16	282.82	263.09	247.82	235.68
	0.7		293.88	258.85	233.91	215.36	201.09	189.81
	0.8			225.00	201.03	183.32	169.76	159.09
	0.9				177.78	160.66	147.62	137.41
	1.0					144.00	131.35	121.48
	1.1						119.01	109.40
	1.2							100.00

Plate thickness mm	Plate width m	0.6	0.7	0.8	0.9	1.0	1.1	1.2
12.00	0.6	576.00	498.07	445.20	407.26	378.86	356.87	339.38
	0.7		423.18	372.74	336.82	310.12	289.57	273.32
	0.8			324.00	289.49	263.97	244.45	229.08
	0.9				256.00	231.35	212.58	197.86
	1.0					207.36	189.15	174.93
	1.1						171.37	157.53
	1.2							144.00
12.5	0.6	625.00	540.44	483.07	441.91	411.09	387.22	368.25
	0.7		459.18	404.45	365.45	336.50	314.21	296.57
	0.8			351.56	314.11	286.43	265.25	248.57
	0.9				277.78	251.03	230.66	214.70
	1.0					225.00	205.24	189.81
	1.1						185.95	170.94
	1.2							156.25

A.21 Relative British Standards and Codes of Practice

CP3	Code of basic data for the design of buildings. Chapter V: Part 2: Wind loads.
CP2004	Foundations.
BS6399	Loading for buildings: Part 1: Code of practice for dead and imposed loads.
BS5950	Structural use of steelwork in building: Part 1: Code of practice for design in simple and continuous construction: hot rolled sections.
BS5950	Part 2: Specification for materials, fabrication and erections: hot rolled section.
BS4	Structural steel sections: Part 1: Hot rolled sections.
BS153	Steel girder bridges.
BS4336	Methods for non-destructive testing of plate material.
BS4360	Weldable structural steels.
BS6072	Method for magnetic particle flaw detection.
BS3923	Methods for ultrasonic examination of welds.
BS4871	Approval testing of welders working to approved welding procedures.
BS5135	Specification for the process of arc welding of carbon and carbon manganese steels.
BS2853	The design and testing of steel overhead runway beams.
BS4190	Isometric black hexagon bolts, screws and nuts.
BS4395	High strength friction grip bolts and associated nuts and washers for structural engineering.
BS3294	The use of high strength friction gap bolts in structural steelwork.
BS4848	Hot rolled structural steel sections: Part 2: Hollow sections; Part 4: Equal and unequal angles.
BS5502	Code of practice for the design of buildings and structures for agriculture.
BS4076	Specification for steel chimneys.
BS5400	Steel, concrete & composite bridges.
BS466	Specification for power driven overhead travelling cranes, semi-goliath and goliath cranes for general use.
BS2573	Rules for the design of cranes.
BS3579	Heavy duty electric overhead travelling and special cranes for use in steel works.
BS5930	Code of practice for site investigations.
BS2655	Part 4: General requirements for escalators and passenger conveyors.
BS8110	Structural use of concrete.

Index